# In Vitro Transcription and Translation Protocols

# METHODS IN MOLECULAR BIOLOGY™

## John M. Walker, SERIES EDITOR

METHODS IN MOLECULAR BIOLOGY™

# In Vitro Transcription and Translation Protocols

## SECOND EDITION

Edited by

## Guido Grandi

*Biochemistry and Molecular Biology Unit,*
*Novartis Vaccines, Siena, Italy*

HUMANA PRESS ✸ TOTOWA, NEW JERSEY

This publication is printed on acid-free paper. ∞
ANSI Z39.48-1984 (American Standards Institute) Permanence of Paper for Printed Library Materials.

Cover design by Nancy K. Fallatt
Cover illustrations: *(background)* Fig. 4A, Chap. 10 (*see* caption on p. 205); *(foreground left)* Fig. 2B, Chap. 6 (*see* caption on p. 115); *(center)* Fig. 2B, Chap. 6 (*see* caption on p. 115); *(right)* Fig. 5, Chap. 6 (*see* caption on p. 125).

For additional copies, pricing for bulk purchases, and/or information about other Humana titles, contact Humana at the above address or at any of the following numbers: Tel.: 973-256-1699; Fax: 973-256-8341; E-mail: orders@humanapr.com; or visit our Website: www.humanapress.com

**Photocopy Authorization Policy:**
Authorization to photocopy items for internal or personal use, or the internal or personal use of specific clients, is granted by Humana Press Inc., provided that the base fee of US $30.00 per copy is paid directly to the Copyright Clearance Center at 222 Rosewood Drive, Danvers, MA 01923. For those organizations that have been granted a photocopy license from the CCC, a separate system of payment has been arranged and is acceptable to Humana Press Inc. The fee code for users of the Transactional Reporting Service is: [978-1-58829-558-3 • 1-58829-558-3/07 $30.00].

Printed in the United States of America. 10 9 8 7 6 5 4 3 2 1
ISBN 10-digit: 1-58829-558-3    ISBN 13-digit: 978-1-58829-558-3
eISBN 10-digit: 1-59745-388-9    eISBN 13-digit: 978-1-59745-388-2
ISSN:1064-3745

**Library of Congress Cataloging-in-Publication Data**

In vitro transcription and translation protocols. -- 2nd ed. / edited by
   Guido Grandi.
      p. ; cm. -- (Methods in molecular biology, ISSN 1064-3745 ;
   v. 375)
   Includes bibliographical references and index.
   ISBN 10-digit: 1-58829-558-3 (alk. paper)
   1. Molecular genetics--Laboratory manuals. 2. Genetic transcription
--Laboratory manuals. 3. Genetic translation--Laboratory manuals.
I. Grandi, Guido. II. Series: Methods in molecular biology (Clifton,
N.J.) ; v. 375.
   [DNLM: 1. Molecular Biology--Laboratory Manuals. 2. Protein
Biosynthesis--Laboratory Manuals. 3. Transcription, Genetic
--Laboratory Manuals. 4. Genetic Techniques--Laboratory Manuals.
W1 ME9616J v.375 2007 / QH 440.5 I62 2007]
QH442.I557 2007
572.8'845078--dc22

                                        2006017634

# Preface

For several years many laboratories have been using in vitro transcription and translation systems as a tool to study protein synthesis and function. In consideration of the general interest in these techniques and with the aim of further encouraging their application in other fields of research, in 1995 Humana Press dedicated volume 37 of the successful *Methods in Molecular Biology* series to publication of detailed protocols for transcription and translation using different in vitro systems.

Over the eleven years since the publication of the first edition the technology of transcription and translation has been improved further and refined to the point that not only can these techniques be adapted to high-throughput screening formats, but they can be also be considered as a competitive alternative to in vivo expression systems for preparative purposes. Indeed, current protein production yields in the mg/mL range, as opposed to the few μg/mL obtainable just a few years ago, allows the exploitation of in vitro protein production methods for several applications in which a fair amount of protein is required.

In consideration of the recent advances in the field, Humana Press now releases the second edition of *In Vitro Transcription and Translation Protocols,* with the expectation that detailed information about newly available techniques will allow laboratories to facilitate and possibly diversify their day-to-day research activities.

*In Vitro Transcription and Translation Protocols* is organized in two parts. The first is dedicated to the presentation of new protocols for in vitro transcription and translation. This includes two chapters describing refined procedures to optimize protein yield using new energy systems for ATP regeneration and an interesting device capable of removing exhausted products and providing the system with additional substrates. Part I also includes a chapter on in vitro production of integral membrane proteins and protocols for high-throughput production of proteins using bacterial and wheat germ extracts. Finally, Part I ends with a detailed description of how *Xenopus* oocytes can be exploited for abundant expression of exogenous proteins.

Part II is dedicated to a collection of chapters in which specific applications of in vitro transcription and translation protocols are described. The applications include the use of in vitro systems in preparative-scale protein production amenable to structural analysis and high-throughput protein production for protein engineering studies and for identification of vaccine candidates.

I do hope that this new edition will be as successful as the previous one in terms of offering the interested scientist new ideas and novel experimental solutions.

*Guido Grandi*

# Contents

# Contributors

FRANK BERNHARD • *Centre for Biomolecular Magnetic Resonance, University of Frankfurt/Main, Institute for Biophysical Chemistry, Frankfurt/Main, Germany*

ELENA BOSSI • *Laboratory of Cellular and Molecular Physiology, Department of Structural and Functional Biology, University of Insubria, Varese, Italy*

KARA A. CALHOUN • *Department of Chemical Engineering, Stanford University, Stanford, CA*

ALDO CERIOTTI • *Institute of Agricultural Biology and Biotechnology, National Research Council, Milan, Italy*

VOLKER DÖTSCH • *Centre for Biomolecular Magnetic Resonance, University of Frankfurt/Main, Institute for Biophysical Chemistry, Frankfurt/Main, Germany*

YAETA ENDO • *Cell-Free Science and Technology Research Center, Ehime University, Matsuyama, Japan*

MARIA SERENA FABBRINI • *Institute of Agricultural Biology and Biotechnology, National Research Council, Milan, Italy*

LAURA FARAVELLI • *Electrophysiology Unit, Newron Pharmaceuticals, Bresso, MI, Italy*

GERMANO FERRARI • *Biochemistry and Molecular Biology Unit, Novartis Vaccines, Siena, Italy*

IGNAZIO GARAGUSO • *Biochemistry and Molecular Biology Unit, Novartis Vaccines, Siena, Italy*

ORIT GAT • *Department of Biochemistry and Molecular Genetics, Israel Institute for Biological Research, Ness-Ziona, Israel*

STEFANO GIOVANNARDI • *Laboratory of Cellular and Molecular Physiology, Department of Structural and Functional Biology, University of Insubria, Varese, Italy*

MUDEPPA D. GOUDA • *Cell-Free Science and Technology Research Center, Ehime University, Matsuyama, Japan*

GUIDO GRANDI • *Biochemistry and Molecular Biology Unit, Novartis Vaccines, Siena, Italy*

HAIM GROSFELD • *Department of Biochemistry and Molecular Genetics, Israel Institute for Biological Research, Ness-Ziona, Israel*

OLIVER HAUSS • *Tumor Genetics Research Group, Max Planck Institute for Molecular Physiology, Dortmund, Germany*

MUTSUNORI IGA • *Division of Infectious Diseases, The Advanced Clinical Research Center, The Institute of Medical Science, The Unviersity of Tokyo, Tokyo, Japan*

CHRISTIAN KLAMMT • *Centre for Biomolecular Magnetic Resonance, University of Frankfurt/Main, Institute for Biophysical Chemistry, Frankfurt/Main, Germany*

YUICHI KOGA • *Department of Material and Life Science, Graduate School of Engineering, Osaka University, Suita Osaka, Japan*

TOSHIYUKI KOHNO • *Molecular Structure Research Group, Mitsubishi Kagaku Institute of Life Sciences (MITILS), Tokyo, Japan*

VYACHESLAV A. KOLB • *Institute of Protein Research, Russian Academy of Sciences, Pushchino, Moscow Region, Russia*

JUN KOMANO • *Laboratory of Virology and Pathogenesis, AIDS Research Center, National Institute of Infectious Diseases, Tokyo, Japan*

AIGAR KOMMER • *Institute of Protein Research, Russian Academy of Sciences, Pushchino, Moscow Region, Russia*

SIMONA MAGAGNIN • *Laboratory of Cellular and Molecular Physiology, Department of Structural and Functional Biology, University of Insubria, Varese, Italy*

ZENE MATSUDA • *Research Center for Asian Infectious Diseases, The Institute of Medical Science, The University of Tokyo, Tokyo, Japan*

KOSUKE MIYAUCHI • *Institute of Human Virology, University of Maryland Biotechnology Institute, Baltimore, MD*

KAZUHIRO MORISHITA • *Department of Tumor and Cellular Biochemistry, Department of Biochemistry, Faculty of Medicine, University of Miyazaki, Miyazaki, Japan*

RYO MORISHITA • *Cell-Free Science, Co., Ltd., Kanagawa, Japan*

OLIVER MÜLLER • *Tumor Genetics Research Group, Max Planck Institute for Molecular Physiology, Dortmund, Germany*

HIDEO NAKANO • *Laboratory of Molecular Biotechnology, Graduate School of Bioagricultural Sciences, Nagoya University, Nagoya, Japan*

NATHALIE NORAIS • *Biochemistry and Molecular Biology Unit, Novartis Vaccines, Siena, Italy*

AKIHIKO OKAYAMA • *Department of Rheumatology, Infectious Diseases, and Laboratory Medicine, Faculty of Medicine, University of Miyazaki, Miyazaki, Japan*

SUANG RUNGPRAGAYPHAN • *Department of Biopharmacy, Faculty of Pharmacy, Silpakorn University, Nakorn-Pathom, Thailand*

TATSUYA SAWASAKI • *Cell-Free Science and Technology Research Center, Ehime University, Matsuyama, Japan*

DANIEL SCHWARZ • *Centre for Biomolecular Magnetic Resonance, University of Frankfurt/Main, Institute for Biophysical Chemistry, Frankfurt/Main, Germany*

AVIGDOR SHAFFERMAN • *Department of Biochemistry and Molecular Genetics, Israel Institute for Biological Research, Ness-Ziona, Israel*

VLADIMIR A. SHIROKOV • *Institute of Protein Research, Russian Academy of Sciences, Pushchino, Moscow Region, Russia*

ANDREA SORAGNA • *Laboratory of Cellular and Molecular Physiology, Department of Structural and Functional Biology, University of Insubria, Varese, Italy*

ALEXANDER S. SPIRIN • *Institute of Protein Research, Russian Academy of Sciences, Pushchino, Moscow Region, Russia*

JAMES R. SWARTZ • *Department of Chemical Engineering, Stanford University, Stanford, CA*

HIROHITO TSUBOUCHI • *Department of Digestive and Life-Style Related Disease, Kagoshima University Graduate School of Medical and Dental Sciences, Kagoshima, Japan*

TSUNEO YAMANE • *Department of Environmental Biology, College of Bioscience and Biotechnology, Chubu University, Kasugai, Japan*

# I

## TECHNOLOGIES

# 1

# Energy Systems for ATP Regeneration in Cell-Free Protein Synthesis Reactions

## Kara A. Calhoun and James R. Swartz

### Summary

Supplying energy for cell-free protein synthesis reactions is one of the biggest challenges to the success of these systems. Oftentimes, short reaction duration is attributed to an unstable energy source. Traditional cell-free reactions use a compound with a high-energy phosphate bond, such as phosphoenolpyruvate, to generate the ATP required to drive transcription and translation. However, recent work has led to better understanding and activation of the complex metabolism that can occur during cell-free reactions. We are now able to generate ATP using energy sources that are less expensive and more stable. These energy sources generally involve multistep enzymatic reactions or recreate entire energy-generating pathways, such as glycolysis and oxidative phosphorylation. We describe the various types of energy sources used in cell-free reactions, give examples of the major classes, and demonstrate protocols for successful use of three recently developed energy systems: PANOxSP, cytomim, and glucose.

**Key Words:** Cell-free protein synthesis; in vitro transcription-translation; energy source; PANOxSP; cytomim; glycolysis.

From: *Methods in Molecular Biology, vol. 375,*
*In Vitro Transcription and Translation Protocols, Second Edition*
Edited by: G. Grandi © Humana Press Inc., Totowa, NJ

## 1. Introduction

Cell-free protein synthesis reactions can be performed in a variety of ways. Extract from different cell sources *(1–3)*, variations in salt or amino acid concentrations *(4,5)*, the addition of specific cofactors or inhibitors *(6,7)*, and different reactor configurations *(8)* have all been used. Regardless of these variations, one common requirement for all cell-free reactions is the supply of ATP to drive the translation or combined transcription and translation processes. The ATP supply is usually generated from a secondary energy source added to the reaction. Traditionally, a compound with a high-energy phosphate bond, such as phosphoenolpyruvate (PEP) *(9)*, creatine phosphate (CP) *(8)*, or acetyl phosphate (AP) *(10)*, is added to the reaction to regenerate ATP in a simple substrate-level phosphorylation reaction. However, these compounds are expensive and can create inhibitory levels of inorganic phosphate in the reaction. Much work has been done in recent years to more fully understand and to further activate the complex metabolism that can occur during cell-free protein synthesis. With more advanced knowledge of and control over cellular metabolism, we no longer have to rely on one-step phosphorylation reactions to drive cell-free protein synthesis. Multistep reactions or entire energy-generating pathways, such as glycolysis and oxidative phosphorylation, have been shown to be active in cell-free reactions *(4,11)*. These energy systems provide ATP from less expensive energy sources while prolonging the protein synthesis reaction. In this chapter, we will discuss *Escherichia coli*-based systems that combine transcription and translation with a specific emphasis on procedures for performing reactions using various energy systems. This chapter supplements another publication in this series by Swartz et al. in 2004 *(12)*.

### 1.1. Energy Requirements During Combined Transcription/Translation Reactions

A constant supply of energy is required to maintain the combined transcription/translation reactions that occur during cell-free protein synthesis. However, the energy requirement for transcription is believed to be negligible relative to translation because mRNA molecules can be used multiple times *(5)*. During translation, energy is required in the form of ATP for acylation of the tRNAs. In addition, GTP is necessary for the

---

**Transcription** – Energy requirement is negligible due to repeated mRNA use (5)

$$nNTP \rightarrow NP_{p,n}(mRNA) + nPP_i$$

**Translation** – Requires 4-5 ATP per amino acid addition

    1.) Charging tRNAs – Requires 2 ATP equivalents

$$AA + ATP \rightarrow AMP\text{-}AA + PP_i$$
$$\underline{AMP\text{-}AA + tRNA \rightarrow tRNA\text{-}AA + AMP}$$
$$AA + ATP \rightarrow tRNA\text{-}AA + AMP + PP_i$$

    2.) Polypeptide synthesis – requires 2-3 ATP equivalents

$$AA_{p,n} + tRNA\text{-}AA + (2\text{-}3)\,GTP \rightarrow AA_{p,n+1} + (2\text{-}3)\,GDP + (2\text{-}3)\,P_i$$

**Energy Requirements for 1 mg/mL protein**

| Protein produced (mg/mL) | Protein Molecular weight (mg/mmol) | Protein produced (mM) | Number of AA per protein | Number of ATP required per AA | ATP required (mM) |
|---|---|---|---|---|---|
| 1 | 25000 | 0.04 | 220 | 4-5 | 35-44 |

Fig. 1

activity of EF-G and EF-Tu complexes that catalyze polymerization of the growing polypeptide chain *(13)*. Assuming four to five ATP equivalents required per amino acid bond, approx 35–44 m*M* of ATP is required to produce 1 mg/mL of an average size protein (25 kD, 220 amino acids) **(Fig. 1)**. This ATP can be generated from a variety of energy sources.

## 1.2. General Considerations When Using an Energy Source

Different secondary energy sources can be added to the cell-free reaction to obtain the necessary ATP. These energy sources contain either a high-energy phosphate bond so that the ATP can be regenerated directly, or they are compounds that can be metabolized through multistep reactions or entire pathways to regenerate nucleoside triphosphates. In choosing a secondary energy source, several issues must be considered including the following: stability, cost, cofactor requirements, scale-up, and effect on protein folding. Energy sources may be unstable in the cell-free reaction where enzymes present in the cell extract can degrade compounds unproductively *(5)*. This has been reported to be a problem in cell-free reactions using compounds with high-energy phosphate bonds that are degraded by nonspecific phosphatases *(14)*. The cost of energy sources is also an important consideration because the energy source is the largest reagent cost in traditional systems, representing almost 50% of the cost *(11)*. When considering alternative energy sources that require multistep reactions to generate ATP, additional compounds are often required to activate pathways. For instance, systems that use pyruvate as an energy source may require the addition of cofactors (nicotinamide adenine dinucleotide [NAD] or coenzyme A [CoA]) *(5)* or exogenous enzymes and oxygen to obtain significant yields *(15)*. Similarly, reactions that require oxygen may face problems meeting the oxygen demand during scale-up. Finally, much progress has been made in folding complex proteins in cell-free systems *(16,17)*. However, the conditions for effective folding may be too harsh for more complicated energy source generation. Cell-free reactions that produce proteins with disulfide bonds, for example, require an oxidizing environment. Unfortunately, stabilization of that oxidizing environment with iodoacetamide pretreatment may inactivate enzymes necessary for energy generation. Overall, when choosing an energy source, controlling energy source stability and cost must be carefully balanced with maintaining appropriate conditions for complex energy-generating pathways and protein folding. In addition, it may be beneficial to inhibit reactions that divert metabolic flux away from energy-generating pathways.

### 1.3. Commonly Used Energy Sources

The reported energy sources used for cell-free reactions can basically be divided into three main classes: compounds with high-energy phosphate bonds for traditional substrate-level phosphorylation, compounds involved in multistep enzymatic pathways, or compounds that help recreate an entire energy-generating pathways. Examples of each class are given next, followed by methods for three of the more complex systems: PANOxSP, cytomim, and glucose.

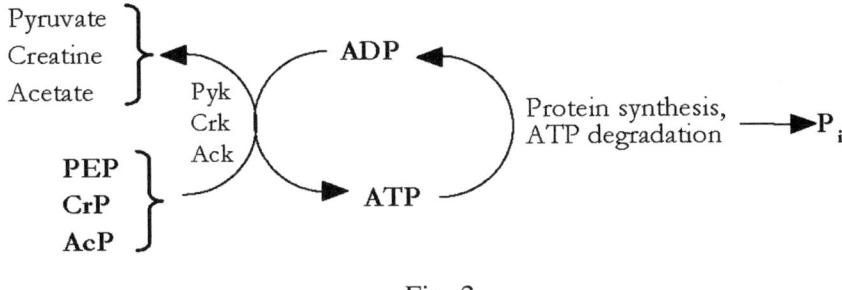

Fig. 2

## 1.3.1. Traditional Substrate-Level Phosphorylation

Traditional batch cell-free reactions contain 30–50 m$M$ of PEP, CP, or AcP to directly phosphorylate ADP to ATP (**Fig. 2**). Although these compounds should be able to provide enough energy to make 1 mg/mL of protein, most reported yields are well below this value. Most likely, the major cause of the low yields is instability of the energy sources as they are degraded by nonspecific phosphatases present in the cell extract *(14)*. Not only does this reduce the productivity of the energy source, but it also generates high concentrations of inorganic phosphate in the cell-free reaction. Inorganic phosphate is known to be inhibitory to protein synthesis at concentrations more than 30 m$M$ *(5)*. To address this instability, reduction of phosphatase activity has been attempted with some success. Kim and Choi suggest that the addition of phosphate in the growth media reduces phosphatase levels in an *E. coli* system *(7)*. The wheat-germ system has been improved through by immunoprecipitation of phosphatases *(14)*. Another method for increasing the supply of the energy source is to use different reactor configurations, such as continuous-flow or semicontinuous reactors *(8)*. These designs allow continued supply of the secondary energy source, while providing a means for inhibitory byproducts (such as inorganic phosphate) to be removed. Finally, the recently described PURE system, which uses all purified components to catalyze protein synthesis has no phosphatase activity *(18)*. Overall, the use of compounds with high-energy phosphate bonds is a simple, yet expensive, way to drive in vitro transcription/translation.

### 1.3.2. Multistep Enzymatic Phosphorylation

Because traditional energy sources are extremely expensive, the use of alternative compounds has been investigated. Most of these alternative energy sources require multistep enzymatic reactions to generate ATP. The pyruvate oxidase system converts pyruvate into AP using an exogenous enzyme *(15)*. Subsequently, the AP phosphorylates ATP in a reaction catalyzed by endogenous acetate kinase (**Fig. 3A**). This system has the advantage of using a less expensive energy source, but does require the supply of molecular oxygen and the addition of an exogenous enzyme. Another system, named PANOxSP (PEP, amino acids, NAD, oxalic acid, spermidine, and putrescine), still uses PEP as the main source of energy, but increases the efficiency of ATP generation through the addition of the cofactors, NAD and CoA, to activate an additional ATP-generating reaction (**Fig. 3B**) *(4)*. In this way, up to 1.5 $M$ of ATP can be obtained from each mole of PEP instead of just 1 $M$-ATP/mole-PEP. The use of oxalic acid is this system is beneficial because it inhibits the activity of PEP synthase *(6)*, an enzyme that wastes ATP by converting pyruvate to PEP. Finally, the use of 3-phosphoglycerate as an energy source has also been reported *(19)*. This compound is converted to PEP by the glycolytic enzymes phosphoglycerate mutase and enolase. This energy source is more stable than PEP alone, resulting in prolonged protein synthesis. By using multistep pathways to generate ATP, these reaction systems have advantages over traditional energy sources in cost, efficiency, and stability.

### 1.3.3. Recreating Entire Energy-Generating Pathways

The ability to activate multistep reactions through addition of cofactors or enzymes suggested the possibility of recreating entire energy-generating pathways in cell-free reactions. For a long time, the metabolic activities catalyzed by the crude cell extract were not well characterized. However, with advanced understanding of this class of cell-free reactions, it is now possible to activate complex metabolic pathways for generating nucleotide triphosphates. The basic hypothesis is that replicating intracellular conditions will activate complex intracellular functions. This concept led to the development of the cytomim system *(4)*. Some of the

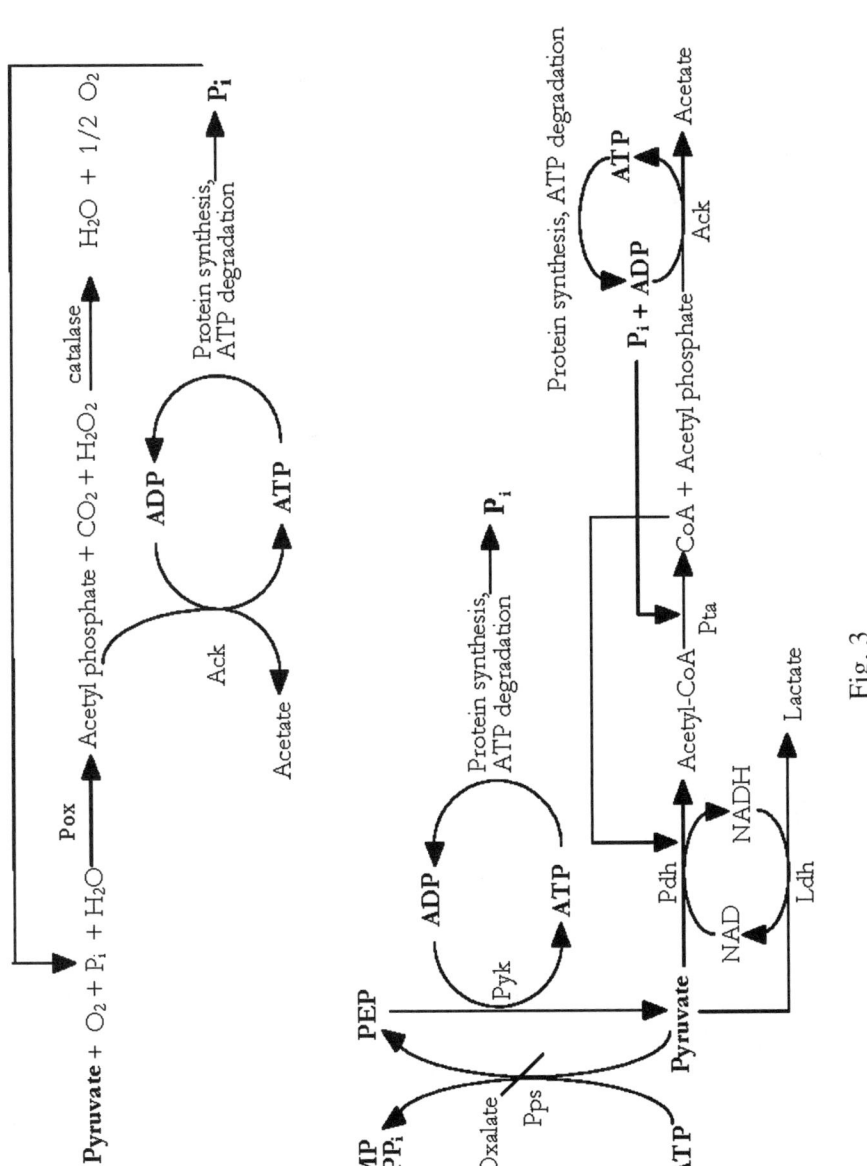

Fig. 3

specific changes include replacing PEG, a high molecular weight compound used for nucleic acid stability, with the natural polyamines putrescine and spermidine. In addition, a nonphosphorylated energy source, pyruvate, replaced PEP, thus avoiding phosphate accumulation during the course of the reaction. This new reaction maintains pH homeostasis, so the buffer can be removed. In addition, acetate salts were replaced with glutamate salts. Together, these changes resulted in protein yields similar to the PANOxSP reactions, except with a much less expensive energy source. Interestingly, the high protein yields are well above what was expected through ATP-generation by conversion of pyruvate to acetate alone. Consequently, another energy-generating pathway, most likely oxidative phosphorylation, is responsible for the additional energy. In fact, the addition of oxidative phosphorylation inhibitors significantly reduces protein yields.

In addition to oxidative phosphorylation, the glycolytic pathway has also been shown to generate energy for cell-free protein synthesis *(11)*. Glucose is the preferred carbon and energy source of many organisms and is also one of the least expensive and most desirable commercial substrates for industrial biotechnology applications. In order for cell-free protein synthesis to effectively compete with conventional in vivo approaches to protein production, it would be highly advantageous to develop a system where glucose can be used as the energy source.

To use glucose for cell-free protein synthesis, the cytomim system had to be adapted slightly. For example, the conversion of glucose to acetate and lactate results in pH instability, so the use of a buffer or other appropriate pH control is necessary. In addition, without the use of a phosphorylated energy source, there is too little inorganic phosphate present in the reaction because phosphate is necessary for the initial steps in the glycolytic pathways. When 90 m$M$ Bis-Tris buffer and 10 m$M$ phosphate are added to cell-free reactions with glucose as the energy source, significant protein yields are possible (~550 µg/mL) *(11)*.

Recreating oxidative phosphorylation and glycolysis in a cell-free environment is important not only for inexpensive ATP generation during protein synthesis, but also as an example of how complex biological systems can be understood and exploited through cell-free biology.

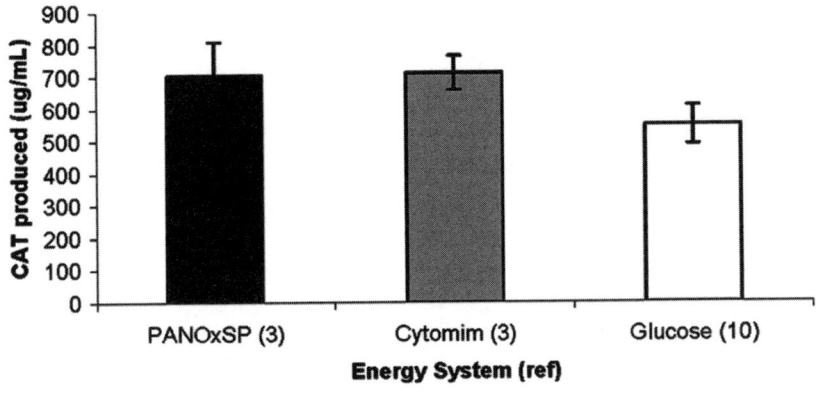

Fig. 4

The following section describes, in detail, the procedure for performing cell-free reactions using the PANOxSP, cytomim, and glucose systems. If these protocols are followed, the total protein yields that can be expected are 700 ± 105, 713 ± 54, and 550 ± 60 μg/mL, respectively **(Fig. 4)**.

## 2. Materials

Unless otherwise stated, all components were purchased from Sigma (St. Louis, MO).

1. 10X Salt solution for PANOxSP system: 200 m*M* magnesium glutamate (Fluka) (*see* **Note 1**), 100 m*M* ammonium glutamate, 175 m*M* potassium glutamate.
2. 10X Salt solution for cytomim and glucose system: 80 m*M* magnesium glutamate (Fluka) (*see* **Note 1**), 100 m*M* ammonium glutamate, 130 m*M* potassium glutamate.
3. 10X Master mix, pH 7.0: 12 m*M* ATP, 8.5 m*M* each of GTP, UTP, and CTP, 340 μg/mL folinic acid, 1.71 mg/mL *E. coli* tRNA (Roche).
4. 50 m*M* Bis-Tris buffer, pH 7.0.
5. 20 Amino acid mix, 50 m*M* each added in the following order (*see* **Note 2**): valine, tryptophan, phenylalanine, isoleucine, leucine, cysteine, methionine, alanine, arginine, asparagine aspartic acid, glutamic acid, glycine, glutamine, histidine, lysine, proline, serine, threonine, tyrosine.

6. 50 m*M* NAD.
7. 50 m*M* CoA.
8. 100 m*M* Putrescine.
9. 100 m*M* Spermidine.
10. 1 *M* PEP (Roche) (*see* **Note 3**).
11. 1 *M* Glucose.
12. 1 *M* Potassium phosphate, pH 7.0.
13. 1 *M* Sodium oxalate.
14. L-[U-$^{14}$C]-leucine (Amersham Pharmacia Biotechnology CFB183).
15. 5 mg/mL T7 RNA Polymerase, prepared as per Swartz et al. *(12)*.
16. DNA template: 1.3 mg/mL pK7CAT plasmid with the *cat* gene coding for chloramphenicol acetyl transferase is used in this protocol for illustration. It is purified using a Qiagen Plasmid Maxi-prep Kit (Valencia, CA).
17. *E. coli* S30 extract, prepared as per Swartz et al. *(12)* (*see* **Note 4**).

## 3. Methods

This protocol describes general considerations for performing cell-free protein synthesis reactions with three different reaction systems: PANOxSP, cytomim, and glucose. Each system has a different secondary energy source. The PANOxSP system is an example of a multistep pathway where cofactors have been added to activate additional pathways. The cytomin system was developed to closely mimic the cellular environment and is believed to derive energy from oxidative phosphorylation. The third system uses glucose, the entire glycolytic pathway, and probably also oxidative phosphorylation to generate ATP for protein synthesis. The final concentrations of the reaction components are listed in **Table 1**. These methods utilize a batch configuration (*see* **Note 5**), and are performed at laboratory scale, usually 15 µL (*see* **Note 6**).

### 3.1. Protein Expression in Various Reaction Systems

1. Prepare stock solutions of materials listed above (*see* **Note 7**).
2. A "premix" is formed by combining the reaction components necessary for three or four replicates of a 15-µL reaction in an Eppendorf tube at 4°C in the order listed above but excluding T7 RNAP, plasmid, and cell extract. After addition of each component, the reagents are mixed by pipetting up and down approx 5–10 times.

**Table 1**
**Reaction Components for Various Cell-Free Protein Synthesis Reactions**

| Solution | Description | Final concentration | | |
|---|---|---|---|---|
| | | PANOxSP | Cytomim | Glucose |
| 1 | Salt solution | 1X | 1X | 1X |
| | Magnesium glutamate (m*M*) | 20 | 8 | 8 |
| | Ammonium glutamate (m*M*) | 10 | 10 | 10 |
| | Potassium glutamate (m*M*) | 175 | 130 | 130 |
| 2 | Master mix | 1X | 1X | 1X |
| | ATP (m*M*) | 1.2 | 1.2 | 1.2 |
| | GTP (m*M*) | 0.85 | 0.85 | 0.85 |
| | CTP (m*M*) | 0.85 | 0.85 | 0.85 |
| | UTP (m*M*) | 0.85 | 0.85 | 0.85 |
| | Folinic acid (mg/mL) | 34 | 34 | 34 |
| | tRNA (mg/mL) | 170.6 | 170.6 | 170.6 |
| 3 | Bis–Tris (m*M*) | 0 | 0 | 50 |
| 4 | 20 Amino acid mix (m*M*) | 2 | 2 | 2 |
| 5 | NAD (mM) | 0.33 | 0.33 | 0.33 |
| 6 | Coenzyme A (m*M*) | 0.27 | 0.27 | 0.27 |
| 7 | Putrescine (m*M*) | 1 | 1 | 1 |
| 8 | Spermidine (m*M*) | 1.5 | 1.5 | 1.5 |
| 9 | PEP (m*M*) | 33 | 0 | 0 |
| 10 | Glucose (m*M*) | 0 | 0 | 33 |
| 11 | Potassium phosphate (m*M*) | 0 | 0 | 10 |
| 12 | Sodium oxalate (m*M*) | 2.7 | 2.7 | 0 |
| 13 | 14C-Leucine (m*M*) | 5 | 5 | 5 |
| 14 | T7 RNA polymerase (mg/mL) | 0.1 | 0.1 | 0.1 |
| 15 | pK7CAT (mg/mL) | 13 | 13 | 13 |
| 16 | *Escherichia coli* S30 extract (% vol) | 0.24 | 0.24 | 0.24 |

3. Autoclaved, Milli-Q water (4°C) is added to the "premix" to bring the reagents to the final concentrations as specified in **Table 1**.
4. The "premix" components are mixed by vortexing. One may centrifuge (~3–5 s) at 4°C to bring down any residual solution on the tube walls. Alternatively, one can mix thoroughly by pipetting up and down.
5. The appropriate volume of cell extract (3.6 μL for a 15-μL reaction) is aliquoted into the bottom of separate Eppendorf tubes where the reaction will take place (the batch reactor). Avoid introducing air bubbles.
6. T7 RNA Polymerase and the DNA template are added sequentially and carefully to the "premix" to avoid introducing air bubbles. Upon each addition, pipet up and down approx 5–10 times (avoid vortexing these components). Once all of the reagents have been combined, mix 10–15 times by pipetting 50% of the total reaction volume. The reagent mixture now contains all of the necessary substrates and salts for the reaction. A separate reagent mixture is also prepared without the plasmid to assess any background protein expression levels.
7. Next, 11.4 μL of the "premix" reagent mixture is added to the Eppendorf tubes that contain extract to obtain the final 15-μL reaction volume. Mix the complete reaction by pipetting approx 5–10 times.
8. The reaction mixture is brought to 37°C and is incubated for 3 h for PANOxSP and glucose reactions and 6 h for the cytomim system.

### *3.2. Evaluating Protein Production by Radioactive Leucine Incorporation*

The total amount of protein produced can be easily assessed using radioactive amino acid incorporation. Activity assays can also be performed, according to the protein that is being produced.

1. To determine protein synthesis yields via [14]C-leucine incorporation, the reaction should be quenched with 100 μL of 0.1 *N* sodium hydroxide (*see* **Note 8**).
2. For each reaction, equal volumes (usually 50 μL) of the quenched reaction mixture can be spotted onto each of two small pieces of Whatman 3MM chromatography paper (cut to fit into a scintillation vial) and dried under a lamp (~45 min).
3. One of each set of the spotted papers will be precipitated with 5% cold tricholoracetic acid (TCA). Place the papers in a small beaker, and add TCA to cover the papers. Allow the TCA to sit for 15 min, drain, and repeat twice for a total of three washes.

4. Add ethanol to the beaker, let sit approx 5 min. Remove ethanol and set papers under lamp to dry.
5. When papers are dry, count the radioactivity with a liquid scintillation counter. Also, count the papers that were not TCA precipitated and washed to get a ratio of precipitated and washed radioactivity vs total and unwashed radioactivity.
6. Protein synthesis yields can be determined from the ratio of washed to unwashed $^{14}C$ counts according to the following equation (*see* **Note 9**):

(ratio washed/unwashed radioactive counts) * (1/no. of leucine residues per protein molecule,13 for CAT) * (Total leucine in reaction, 2005 µ*M*) * (Molecular weight of protein, 25,662 Da for CAT) / 1000 = amt of protein synthesized (µg/mL)

## 4. Notes

1. The final magnesium concentration in the reaction should be optimized for each batch of extract. Synthesis with a model protein should be performed over a range of magnesium concentrations, typically 12–20 m*M* for PANOxSP reactions and 6–12 m*M* for cytomim and glucose reactions.
2. Use amino acids with more than 98% purity. The order of addition seeks to minimize the white precipitate that forms in the mixture. However, owing to the relatively low solubility of tyrosine, it is important to mix the solution thoroughly before addition to the cell-free reaction.
3. We have seen variability in the cell-free reaction depending on the source of PEP. Our best results come when using the PEP supplied by Roche.
4. The method described by Swartz et al. *(12)* uses complex media in a low-density fermentation to grow the cells used for extract preparation. Alternatively, recently developed extract preparation procedures have also been described without any change in extract performance. For instance, defined media in a moderate-density fermentation can be used, as described by Zawada and Swartz *(20)*. In addition, Liu et al. *(21)* investigated streamlining the extract preparation procedure to reduce time and cost.
5. Cell-free reactions can be performed in several types of reactors. Typically, the batch format has been used owing to its simplicity and ease of operation. Continuous and semicontinuous systems *(8)*, which involve the continuous supply of substrates and removal of inhibitory byproducts, have also shown success. These reactions increase the protein yield and duration of the translation reaction as compared to the batch system.
6. Cell-free reactions are typically less productive when the scale is increased. The best way to increase scale is to maintain high surface-volume

ratio, possibly by spreading a drop of the cell-free reaction in the bottom of a six-well plate (100–250 µL) or a Petri dish (1 mL or more).

7. The radioactive leucine (**Subheading 2.**, **step 13**) is included to determine protein synthesis yields via TCA precipitation followed by liquid scintillation counting. This component can be removed if protein yields are determined another way.

8. To assess the amount of total protein produced, directly quench the cell-free reaction with sodium hydroxide. The soluble amount of protein produced can also be determined in this way by centrifuging the cell-free sample for 15 min at 14,000$g$ (4°C). Then the supernatant can be added to the sodium hydroxide and the protocol continues as usual.

9. Radioactive amino acid incorporation measurements determine the amount of protein that is precipitated onto the filter paper. Thus, all the measured protein does not have to be full-length. However, in our experience, the majority of the protein produced is full-length.

## References

1. Jewett, M. C., Voloshin, A., and Swartz, J. (2002) in *Gene Cloning and Expression Technologies,* (Weiner, M. and Lu, Q., eds.), Eaton Publishing, Westborough, MA, pp. 391–411.
2. Yao, S. -L., Shen, X. -C., and Suzuki, E. (1997) *J. Ferment. Bioeng.* **84,** 7–13.
3. Michel-Reydellet, N., Calhoun, K., and Swartz, J. (2004) Amino acid stabilization for cell-free protein synthesis by modification of the *Escherichia coli* genome. *Metab. Eng.* **6,** 197–203.
4. Jewett, M. C. and Swartz, J. R. (2004) Mimicking the *Escherichia coli* cytoplasmic environment activates long-lived and efficient cell-free protein synthesis. *Biotechnol. Bioeng.* **86,** 19–26.
5. Kim, D. M. and Swartz, J. R. (2001) Regeneration of adenosine triphosphate from glycolytic intermediates for cell-free protein synthesis. *Biotechnol. Bioeng.* **74,** 309–316.
6. Kim, D. -M. and Swartz, J. R. (2000) *Biotechnol. Lett.* **22,** 1537–1542.
7. Kim, R. G., and Choi, C. Y. (2000) Expression-independent consumption of substrates in cell-free expression system from *Escherichia coli. J. Biotechnol.* **84,** 27–32.
8. Spirin, A. S., Baranov, V. I., Ryabova, L. A., Ovodov, S. Y., and Alakhov, Y. B. (1988) A continuous cell-free translation system capable of producing polypeptides in high yield. *Science* **242,** 1162–1164.

9. Kim, D. M., Kigawa, T., Choi, C. Y., and Yokoyama, S. (1996) A highly efficient cell-free protein synthesis system from *Escherichia coli*. *Eur. J. Biochem.* **239,** 881–886.
10. Ryabova, L. A., Vinokurov, L. M., Shekhovtsova, E. A., Alakhov, Y. B., and Spirin, A. S. (1995) Acetyl phosphate as an energy source for bacterial cell-free translation systems. *Anal. Biochem.* **226,** 184–186.
11. Calhoun, K., and Swartz, J. (2005) Energizing cell-free protein synthesis with glucose metabolism. *Biotechnol. Bioeng.* **90,** 606–613.
12. Swartz, J., Jewett, M. C., and Woodrow, K. (2004) in *Recombinant Gene Expression: Reviews and Protocols,* (Balbas, P. and Lorence, A., eds.), Humana Press, Totowa, NJ, pp. 169–182.
13. Stryer, L. (1995) *Biochemistry.* W. H. Freeman and Company, New York.
14. Shen, X.-C., Yao, S.-L., Terada, S., Nagamune, T., and Suzuki, E. (1998) *Biochem. Eng. J.* **2,** 23–28.
15. Kim, D. M., and Swartz, J. R. (1999) Prolonging cell-free protein synthesis with a novel ATP regeneration system. *Biotechnol. Bioeng.* **66,** 180–188.
16. Yin, G., and Swartz, J. R. (2004) Enhancing multiple disulfide bonded protein folding in a cell-free system. *Biotechnol. Bioeng.* **86,** 188–195.
17. Kim, D. M., and Swartz, J. R. (2004) Efficient production of a bioactive, multiple disulfide-bonded protein using modified extracts of *Escherichia coli*. *Biotechnol. Bioeng.* **85,** 122–129.
18. Shimizu, Y., Inoue, A., Tomari, Y., et al. (2001) Cell-free translation reconstituted with purified components. *Nat. Biotechnol.* **19,** 751–755.
19. Sitaraman, K., Esposito, D., Klarmann, G., Le Grice, S. F., Hartley, J. L., and Chatterjee, D. K. (2004) A novel cell-free protein synthesis system. *J. Biotechnol.* **110,** 257–263.
20. Zawada, J. and Swartz, J. (2005) *Biotechnol. Bioeng.* **89,** 407–415.
21. Liu, D. V., Zawada, J., and Swartz, J. (in press) *Biotechnol. Prog.*

# 2

# Continuous-Exchange Protein-Synthesizing Systems

## Vladimir A. Shirokov, Aigar Kommer, Vyacheslav A. Kolb, and Alexander S. Spirin

### Summary

Protein synthesis in cell-free systems is an emerging technology already competing with in vivo expression methods. In this chapter the basic principles of continuous-exchange protein synthesizing systems, and protocols for *Escherichia coli* and wheat germ translation and transcription-translation systems are described. The ways to improve substrate supply in cell-free systems and mRNA design for eukaryotic system are discussed. Correct folding of the synthesized protein is demonstrated and discussed in detail.

**Key Words:** Continuous-exchange cell-free (CECF) system; *E. coli* extract; wheat germ extract; translation reaction mixture; combined transcription-translation; 5'-untranslated region (5'-UTR); 3'-untranslated region (3'-UTR); protein folding; molecular chaperones; nonionic detergents.

## 1. Introduction

The methodology of cell-free protein synthesis in a continuous mode was first proposed in the form of the continuous-flow cell-free (CFCF) systems: the feeding of the synthesis reaction with substrates and the removal

From: *Methods in Molecular Biology,* vol. 375,
*In Vitro Transcription and Translation Protocols, Second Edition*
Edited by: G. Grandi © Humana Press Inc., Totowa, NJ

of products was achieved by the continuous pumping of a feeding solution with amino acids and NTPs into the reactor and the pumping-out of the same volume through an ultrafiltration membrane *(1–4)*. The porous barrier retained ribosomes and other high-molecular-weight components of the protein-synthesizing machinery within a defined reaction compartment, and the continuous feeding made the synthesis sustainable and productive. At the same time it was mentioned that "the simplest configuration (of the reactor) is a membrane bag containing the reaction region, while retaining a solution outside the membrane, which provides for the desired level of the lower molecular weight components in the reaction region. Thus, by exchange across the membrane, the lower molecular weight products produced by the reaction will be continuously dialyzed into the external solution, while the reaction components will be continuously replenished in the reaction region" *(4)*. Because of its simplicity, it is the dialysis format, or continuous-exchange cell-free (CECF) system (*see* **refs. 5–8**), that was reproduced in a number of other laboratories *(9–14)* and became most popular as "a high-throughput protein production method based on cell-free system" (*[15]; see also* **refs. 16** and *17*). Both prokaryotic (*Escherichia coli*) and eukaryotic (wheat germ) cell free translation systems are capable of synthesizing up to several milligrams of protein per milliliter of reaction mixture in CECF format. The *E. coli* system has high translation rate and reach this yield in an overnight CECF run. In the wheat germ system the synthesis rate is lower but owing to higher stability of this system it can work for several days, reaching the total yield of the same range. The cell-free translation systems based on the extracts of eukaryotic cells seems to be of particular interest for the syntheses of eukaryotic proteins. The *E. coli* system works in a transcription-translation mode with DNA template. Wheat germ system can be used in translation mode with RNA template or in transcription-translation mode. For detailed discussion of CECF and other cell-free expression systems see the recent reviews *(7,8,14,15)*. The commercial versions of the high yield CECF expression systems based on *E. coli* or wheat germ extracts, the Rapid Translation Systems (RTS), are available from Roche Diagnostics GmbH, Germany (*[13,17]*; www.protein expression.com).

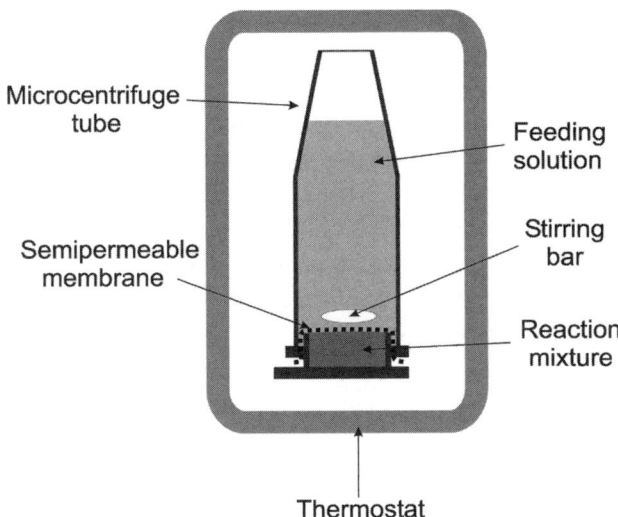

Fig. 1. Cap continuous-exchange cell-free microreactor made from 0.65-mL microcentrifuge tube.

## 2. Materials

### 2.1. CECF Reactors

Simple disposable "cap microreactor" for CECF protein synthesis can be made from microcentrifuge tube of 0.65 mL volume **(Fig. 1)**. The reactor is suitable for analytical or microscale protein production in 50–80 μL of the reaction mixture, yielding 50–200 μg of the protein product. Alternatively, the commercial dialyzers (e.g., Spectra/Por Dispo-Dialyzers from Spectrum Laboratories) can be used for reaction volumes from 0.05 to 1 mL. Unlike the cap microreactor DispoDialyzers provide an access to the reaction mixture during the run (*see* **Note 1**).

#### 2.1.1. Reactor Components

1. 0.65-mL Microcentrifuge tubes. The actual reaction volume may vary for tubes of different manufacturers. Determine the cap volume beforehand by filling it with water until small meniscus appears over the rim: that is the volume of the reaction mixture to be prepared for one sample. The

smaller cap volume gives more favorable ratio between the feeding solution and reaction mixture volumes.

2. Dialysis membrane, regenerated cellulose, 10–15 kDa cut-off. Dialysis membranes made of regenerated cellulose (RC) 20–30 μm in thickness are recommended (e.g., Visking dialysis membrane (Medicell International, UK). Membranes made of cellulose ester (CE) are rigid and may crack during reactor assembly. The RC membrane must be pretreated before use by incubation in a large volume of 2% sodium bicarbonate and 1 mM EDTA at 80°C for 30 min. Then the membrane is rinsed thoroughly in distilled water and stored in 24% ethanol at 4°C.

3. Magnetic stirring bars, 2 × 5 mm, Teflon-coated. The bar surface must be smooth.

### 2.1.2. Preparation of the "Cap Microreactor"

1. Cut off the strip connecting the cap and the body of sterile 0.65-mL microcentrifuge tube. Cut off the very bottom of the tube body to make opening of 3–4 mm in diameter. Keep caps and tubes in a Petri dish.

2. Cut out the 10 mm diameter discs of the pretreated dialysis membrane (regenerated cellulose, 10–15 kDa cut-off) using the cork borer. Rinse membrane discs with distilled water and boil in water for 5 min.

## 2.2. E. coli Transcription-Translation CECF System

### 2.2.1. Solutions and Buffers

Reagents should be DNase-, RNase-, and protease-free. Sterilize by filtration through 0.22-μm filter. Store frozen at –80°C.

1. NTP solution: 90 mM ATP, 60 mM GTP, 60 mM UTP, 60 mM CTP (adjust pH to 7.0 with KOH).
2. 100 mM Glutathione reduced (GSH).
3. 100 mM Glutathione oxidized (GSSG).
4. 10 mg/mL Folinic acid.
5. 1 M Phosphoenol pyruvate K-salt (PEP) (adjust pH to 7.0 with KOH).
6. 1 M Acetyl phosphate K-Li salt (AcP).
7. 19 Amino acids without leucine (Aa's-Leu), 4 mM each.
8. 100 mM Leucine (Leu).
9. Arg, Cys, Trp, Met, Asp, Glu mixture (RCWMDE), 16.7 mM each.
10. 2.5 M HEPES-KOH, pH 8.0.

11. 5 $M$ KOAc.
12. 0.5 $M$ Mg(OAc)$_2$.
13. 40% Polyethyleneglycol 8000 (PEG 8000).
14. 3% NaN$_3$.
15. 40 mg/mL Total tRNA from *E. coli* MRE 600 (tRNA).
16. Complete® protease inhibitors cocktail mini (Complete), 1 tablet per 100 mL (Roche Diagnostics GmbH, Germany).
17. Buffer A: 10 m$M$ Tris-acetate, pH 8.2, 14 m$M$ Mg(OAc)$_2$, 60 m$M$ KCl, 6 m$M$ β-mercaptoethanol.
18. Buffer B: 10 m$M$ Tris-acetate, pH 8.2, 14 m$M$ Mg(OAc)$_2$, 60 m$M$ KCl, 1 m$M$ DTT.
19. Buffer C: 10 m$M$ Tris-acetate, pH 8.2, 14 m$M$ Mg(OAc)$_2$, 60 m$M$ KOAc, 0.5 m$M$ DTT.
20. S-30 buffer: 25 m$M$ HEPES-KOH, pH 8.0, 14 m$M$ Mg(OAc)$_2$, 60 m$M$ KOAc, 0.5 m$M$ DTT.
21. Run-off mixture: 0.75 $M$ Tris-acetate pH 8.2, 7.5 m$M$ DTT, 21 m$M$ Mg(OAc)$_2$, 0.5 m$M$ each of the 20 amino acids, 6 m$M$ ATP, 0.5 $M$ acetylphosphate.
22. Bidistilled or Milli-Q water (H$_2$O).

## 2.2.2. Enzymes

1. T7 bacteriophage DNA-dependent RNA polymerase, 100 U/μL (Roche Diagnostics GmbH, Germany).
2. Human placenta ribonuclease inhibitor (HPRI), 40 U/μL (Roche Diagnostics GmbH).
3. Pyruvate kinase (PK), 10 mg/mL (specific activity, 200 U/mg), solution containing 50% glycerol (Roche Diagnostics GmbH).

## 2.2.4. Nucleic Acids, Vectors, and Bacterial Strains

1. *E. coli* K12, strain A19 is used for preparation of S30 extract (*see* **Note 2**).
2. pET vector or equivalent. For the many variations of this plasmid series, please refer to the Novagen catalog (or www.novagen.com) (*see* **Note 3**).
3. Prepare plasmid DNA by CsCl method (*18*), or with ion exchange column kit (Qiagen or Macherey-Nagel) followed by phenol-chloroform deproteinization. Keep plasmid solution frozen at –20°C for long time storage or at +4°C for everyday use.

## 2.2.5. Equipment

1. Reactor for CECF translation: microreactor made from 0.65-mL micro-centrifuge tube or dialysis reactor of other type.
2. Thermostat-controlled box to maintain the temperature of the reactor (the reactor is placed into the box).
3. Magnetic stirrer.
4. Fermenter (10 L).
5. French press or equivalent instrument.

## 2.3. Wheat Germ Translation and Transcription-Translation CECF Systems

### 2.3.1. Solutions

Sterilize solutions by filtration through 0.22-μ filter. Store at –20°C.
1. 1.0 $M$ HEPES-KOH pH 8.0.
2. 100 m$M$ Mg(OAc)$_2$.
3. 2.5 $M$ KOAc.
4. 3% NaN$_3$.
5. 40% Glycerol.
6. 100 m$M$ Spermidine.
7. 500 m$M$ DTT.
8. 500 m$M$ Creatine phosphate.
9. 200 m$M$ GTP, pH 7.0.
10. 500 m$M$ ATP, pH 7.0.
11. 19 Amino acids mixture without leucine, 3 m$M$ each.
12. 20 m$M$ Leucine.
13. 2 mg/mL Total yeast tRNA.
14. Milli-Q water.

### 2.3.2. Enzymes and Templates

1. Creatine kinase (Roche Diagnostics, Germany), 10 mg/mL solution in 40 m$M$ potassium phosphate, pH 7.5, 10 m$M$ 2-mercaptoethanol, 50% glycerol.
2. Human placenta RNase inhibitor (HPRI), 40 U/μL (Roche Diagnostics).
3. T7 bacteriophage DNA dependent RNA polymerase, 100 U/μL.

4. mRNA encoding the gene sequence with 5'- and 3'-untranslated regions (UTRs) (*see* **Note 3**). Prepare mRNA by in vitro transcription as described in **ref. *19***.
5. Plasmid containing the target gene and sequences of T7 polymerase promoter and 5'- and 3'-UTRs
6. PCR fragment containing the target gene and sequences of T7 polymerase promoter and 5'- and 3'-UTRs.

## 2.3.3. Premixed Solutions

1. Extraction buffer: 80 m*M* HEPES-KOH pH 7.6, 2 m*M* Mg(OAc)$_2$, 4 m*M* CaCl$_2$, 100 m*M* KOAc, 8 m*M* DTT.
2. WGE buffer: 40 m*M* HEPES-KOH pH 7.6, 5 m*M* Mg(OAc)$_2$, 100 m*M* KOAc, 4 m*M* DTT, 0.3 m*M* each 19 Aa's-Leu.
3. Concentrated translation pre-MIX.

   Combine the following components to prepare 20-fold concentrated pre-MIX:

   | | |
   |---|---|
   | H$_2$O | 465 µL |
   | 1.0 *M* HEPES-KOH pH 8.0 | 750 µL |
   | 500 m*M* DTT | 90 µL |
   | 100 m*M* Spermidine | 75 µL |
   | 500 m*M* ATP | 60 µL |
   | 200 m*M* GTP | 60 µL |

   Total volume 1.5 mL is sufficient for approx 50 CECF reactions.

4. Concentrated transcription-translation pre-MIX.
   Combine the following components to prepare 20-fold concentrated pre-MIX:

   | | |
   |---|---|
   | H$_2$O | 285 µL |
   | 1.0 *M* HEPES-KOH pH 8.0 | 750 µL |
   | 500 m*M* DTT | 90 µL |
   | 100 m*M* Spermidine | 135 µL |
   | 500 m*M* ATP | 60 µL |
   | 200 m*M* GTP | 60 µL |
   | 200 m*M* CTP | 60 µL |
   | 200 m*M* UTP | 60 µL |

   Total volume 1.5 mL is sufficient for approx 50 CECF reactions.

   Store at −20°C in aliquots. May be refrozen several times.

# 3. Methods

## 3.1. Protein Synthesis in E. coli Transcription-Translation CECF System

### 3.1.1. PREPARATION OF THE E. COLI S-30 EXTRACT

S30 extract is prepared according to Zubay's procedure with minor modifications *(20)*.

1. Grow *E. coli* strain A19 cells (*see* **Note 2**) in a fermenter in LB medium at 37°C with intensive aeration up to OD 0.8 at 590 nm. Stop the fermentation by fast cooling of the cell culture to 10°C in 5–10 min (*see* **Note 4**), and collect cells by centrifugation in a JA-14 Beckman rotor at 10,000*g* and 4°C for 30 min.
2. Resuspend the collected cells in an equal volume of buffer A and centrifuged for 30 min at 10,000*g*. Discard the supernatant and repeat washing. Freeze cell paste (in form of thin plate) in a liquid nitrogen and store at −70°C overnight.
3. In the morning thaw the pellet and resuspend it in 10-fold volume of buffer A. Centrifuge the cell suspension for 30 min at 10,000*g*. Resuspend the cell pellet in equal volume of buffer B.
4. Add 0.1 m*M* phenylmethyl sulfonyl fluoride (PMSF). Disintegrate cells by passing suspension through a cooled French press or equivalent instrument. (Ultrasonic disintegrators are not suitable.) Keep the homogenate on ice.
5. Centrifuge the homogenate for 30 min at 30,000*g*, collect the upper two-thirds of supernatant, recentrifuge at the same conditions and again collect the upper two-thirds of supernatant.
6. To every 10 mL of supernatant add 0.7 mL of run-off mixture. Incubate at 37°C for 80 min.
7. After the incubation dialyze the mixture against buffer C. Change the buffer after 2 h and continue dialysis overnight.
8. Finally clarify the S30 extract by centrifugation at 10,000*g*. Freeze the extract in portions in liquid nitrogen and store at −80°C.

## 3.1.2. In Vitro Synthesis of Proteins

The following protocol is appropriate for the CECF expression of the plasmid carrying the human proinsulin sequence under the control of the

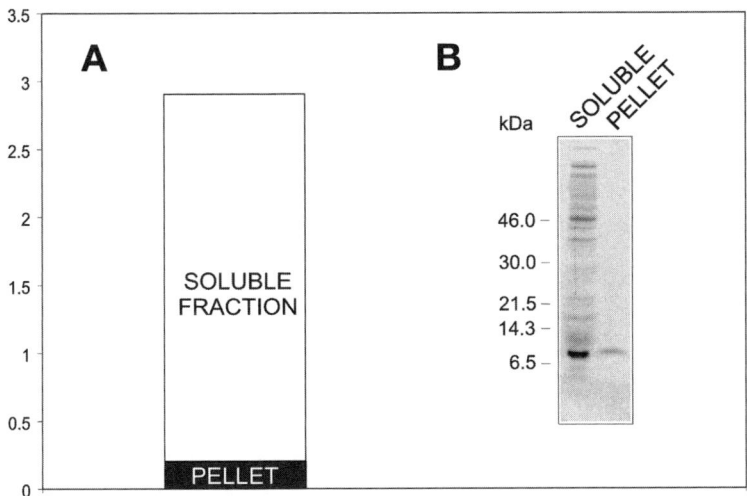

Fig. 2. Synthesis of human proinsulin in *Escherichia coli* continuous-exchange cell-free transcription-translation system. (**A**) Amount of synthesized protein determined by radioactivity of [$^{14}$C]Leucine, incorporated in either soluble protein (white bar) or aggregated product (black bar). (**B**) SDS-PAGE analysis of the synthesized protein, staining with Coomassie G-250. Plasmid pETinsHis6 carrying a full-length copy of human proinsulin cloned into a pET20b(+) vector (Novagen) was kindly provided by Dr. R. S. Esipov (*see* **ref. 51**).

T7 RNA polymerase promoter (**Fig. 2**). The reactor is made from 0.65-mL microcentrifuge tube (**Fig. 1**). Any tube from any manufacturer can be used (however the inner volume of the tube cap may vary, *see* **Subheading 2.2.1.**). Any commercial microdialyser can also be used. No significant difference was observed in the productivity of different reaction volumes recalculated per milliliter. The ratio of reaction mixture/feeding mixture volumes is 1:10.

3.1.2.1. PREPARATION OF FEEDING SOLUTION STOCK

Thaw stock solutions just before starting the protocol. Vortex and store the thawed solutions on ice. For every 100 µL of reaction mixture + 1 mL of feeding solution, mix in the following order in a sterile microcentrifuge tube:

| | |
|---|---|
| $H_2O$ | 57.0 μL |
| 2.5 *M* HEPES-KOH pH 8.0 | 44.0 μL |
| 0.5 *M* Mg(OAc)$_2$ | 16.5 μL |
| 5 *M* KOAc | 33.0 μL |
| 1 *M* AcP | 22.0 μL |
| 1 *M* PEP | 22.0 μL |
| 4 m*M* Aa's-Leu | 137.5 μL |
| 100 m*M* Leu | 5.5 μL |
| 16.7 m*M* RCWMDE | 66.0 μL |
| 10 mg/mL Folinic acid | 11.0 μL |
| NTP solution | 14.7 μL |
| 100X Complete solution | 11.0 μL |
| 3% NaN$_3$ | 22.0 μL |
| 40% PEG 8000 | 55.0 μL |
| 100 m*M* GSH | 11.0 μL |
| 100 m*M* GSSG | 22.0 μL |

Total volume of the feeding solution stock is 550 μL. Take 1/11th of this volume [50 μL] in a separate tube in which the reaction mixture will be prepared.

### 3.1.2.2. PREPARATION OF FEEDING SOLUTION

Mix in the following order:

| | |
|---|---|
| Feeding solution stock | 500 μL (the rest of solution prepared in **Subheading 3.1.2.1.**) |
| S-30 buffer | 175 μL |
| $H_2O$ | 325 μL |

Total volume of the feeding solution is 1 mL. Place solution in thermostat at 30°C.

### 3.1.2.3. PREPARATION OF REACTION MIXTURE

1. Centrifuge the S30 extract of *E. coli* at maximum speed for 10 min at 4ºC in a microcentrifuge. Collect the supernatant and transfer it to a fresh tube.

**Table 1**
**Composition of the Reaction Mixture for *E. coli* Transcription-Translation CECF System**

| Component | Final concentration | Component | Final concentration |
|---|---|---|---|
| S30 extract of *E. coli* | [ 17.5 vol % ] | 14 Aa's (-RCWMDE), each | 0.5 m$M$ |
| pETinsHis$_6$ | [ 0.04 mg/mL ] | RCWMDE, each | 1.5 m$M$ |
| T7 RNA polymerase | [ 2 U/μL ] | Folinic acid | 0.1 m$M$ |
| Total tRNA *E. coli* | 0.5 mg/mL | ATP | 1.2 m$M$ |
| Pyruvate kinase | 0.02 mg/mL | GTP | 0.8 m$M$ |
| HPRI | 0.5 U/μL | CTP | 0.8 m$M$ |
|  |  | UTP | 0.8 m$M$ |
| HEPES-KOH, pH 8.0 | 100 m$M$ | Complete® | 1X |
| Mg(OAc)$_2$ | [10 m$M$*] | NaN$_3$ | 0.05% |
| KOAc | [160 m$M$*] | PEG 8000 | 2% |
| AcP | 20 m$M$ | GSH | [1 m$M$ ] |
| PEP | 20 m$M$ | GSSG | [2 m$M$ ] |

Concentrations marked with asterisk are calculated as a sum of increments added as individual component and as a part of S30 solution (e.g., 7.5 mM Mg[OAc]$_2$ is added as the salt solution, another 2.5 mM comes with 17.5% [v/v] of S30 extract containing 14 mM Mg[OAc]$_2$). Concentrations in brackets should be optimized for particular batch of S30 and gene/plasmid construct. Feeding solution contains only the components shown in gray area.

2. Prepare the transcription-translation reaction mixture. For each 100 µL mix in the following order:

| | |
|---|---|
| Feeding concentrate | 50.0 µL |
| $H_2O$ | 27.0 µL |
| pETinsHis$_6$, 5.5 mg/mL | 0.7 µL |
| HPRI, 40 U/µL | 1.3 µL |
| T7 RNA polymerase, 100 U/µL | 2.0 mL |
| Total tRNA from *E. coli*, 40 mg/mL | 1.3 µL |
| Pyruvate kinase, 10 mg/mL | 0.2 µL |
| S30 extract of *E. coli* | 17.5 µL |
| Total volume is 100 µL | |

The **Table 1** provides standard reaction conditions for *E. coli* transcription translation.

## 3.1.2.4. *E. coli* TRANSCRIPTION-TRANSLATION CECF PROCEDURE

1. Place predetermined volume (*see* **Subheading 2.1.1.**, **step 1**) of the reaction mixture into the tube cap. Using thin forceps take the membrane circle and remove excess water by blotting the membrane edge over the filter paper. Cover the cap with membrane circle placing it nearly flat on the meniscus, center the membrane on the cap. Avoid trapping the bubble under the membrane.
2. Close the cap with tube body: position the tube vertically over the cap and press it straight down firmly without displacing the membrane.
3. Fill the reservoir (tube) with 0.5 mL of feeding solution, insert the magnetic bar and seal the opening with a Parafilm strip. Use another Parafilm strip to seal the contact between a cap and a tube to prevent drying through the membrane edge.
4. Place the reactor(s) in a 30°C thermostated box on a magnetic stirrer. Incubate for 12 h (overnight) with stirrer rotation speed 120–150 rpm.
5. Discard feeding solution. Transfer reaction mixture to a fresh microcentrifuge tube. Centrifuge the mixture at maximum speed for 10 min at room temperature in a microcentrifuge. Collect the supernatant and transfer it to a fresh tube. Reconstitute the pellet in the same volume of S30 buffer or Tris-buffered 8 *M* urea.

   Analyze supernatant and pellet fractions with sodium dodecyl sulfate-polyacrylamide gels (SDS-PAGE). Stain the gel with Coomassie Brilliant Blue or carry out an immunoblot to visualize the synthesized protein. Determine the product concentration by a suitable product activity test (soluble product).

## *3.2. Protein Synthesis in CECF Systems With Wheat Germ Extract*

### *3.2.1. Preparation of Wheat Embryos*

1. Grind the wheat seeds in a blender type coffee mill to release the embryos (*see* **Note 5**).
2. Use three to four grinding pulses of 1–2 s each for every portion of seeds. Process about 1 kg of seeds (that will finally give 2–3 g of embryos).
3. Fractionate the material on a pair of sieves with appropriate mesh sizes to select the fraction containing the embryos.
4. (Work under fume hood!) Separate the embryos from endosperm fragments by flotation in organic solvents: pour the sieved material into the mixture of carbon tetrachloride:cyclohexane (600:260), density 1.354 g/cm), mix and observe separation. The embryos should float while the endosperm fragments sediment. If necessary adjust the density by addition of the appropriate solvent to get better separation. Collect the floating embryos and let them dry overnight under the fume hood.
5. Inspect the material using the big magnifier lens and remove damaged embryos and all the remaining fragments of endosperm or seed envelope (*see* **Note 6**). Weigh up the embryos to determine the amount of extraction buffer (*see* **Subheading 3.2.2**, **step 5**).

### *3.2.2. Preparation of Wheat Germ Extract*

The protocol is based on the procedure of Erickson and Blobel *(21)* with the modifications suggested by Madin et al. *(12)*. The modifications include embryo-washing step, which serves to clean up the embryos from the endosperm contaminants.

Make all the steps at +4°C (in a cold room) with cold solutions. Proceed through **step 1** without delay.

1. Stir embryos (2 g) in 100 mL of ice cold water for 3 min. Do not use magnetic stirring bar as it can damage the embryos. Use mechanical or manual stirring. Discard water and repeat washing step two times more. Then place embryos in a cold 0.5% solution of NP-40 and sonicate them for 3 min in ultrasonic cleaner (e.g., Transsonic T310, Elma, Germany). Discard NP-40 solution, rinse embryos with 100 mL of cold water and sonicate again in 100 mL water for 3 min. Rinse with 200 mL of cold water on a filter, finally remove excess water with suction. The weight of embryos increases by a factor of 2–2.5 as they swell in water.
2. Immediately transfer the washed embryos into liquid nitrogen ($LN_2$) in the prefrozen mortar. (Prefreeze the mortar and pestle at –80°C overnight and then with $LN_2$.)

3. Grind the embryos with a pestle for 10 min to make a thin powder. Add $LN_2$ intermittently to keep the powder wet in $LN_2$. If larger amount of embryos are processed at once, store them under $LN_2$ and grind in 2-g portions.

4. Transfer the powdered embryos to another (cold but not frozen) mortar and let them thaw in a cold room for 10–20 min.

5. Add the extraction buffer in amount equal to the initial weight of embryos (*see* **step 6** in **Subheading 3.2.1.**) and grind the paste thoroughly with a pestle for 5 min. The paste volume will be three to three and a half times the volume of the added buffer.

6. Transfer the paste into centrifuge tubes and spin for 30 min at 30,000$g$ (18,000 rpm in 50Ti rotor, Beckman).

7. Move the floating lipid layer aside and collect the liquid phase (2–2.5 mL). Avoid contamination from both pellet and upper layer.

8. Apply the extract onto PD-10 column (Amersham-Pharmacia Biotech) equilibrated in WGE buffer. After sample absorption add 3 mL of WGE buffer to the column and collect the eluate (*see also* the PD-10 Manual) in 0.5-mL fractions in Eppendorf tubes. Combine five fractions containing milky-white solution. If larger amount of embryos are processed at once, use several PD-10 columns in parallel.

9. Centrifuge the extract at 14,000$g$ for 10 min. Collect supernatant, distribute in aliquots and freeze in $LN_2$. Determine the $A_{260}$ of the extract; it should be at least 200 OU/mL. Store at $-80°C$ or lower. Avoid refreezing, use only fresh aliquots of the extract for setting the CECF reaction.

## 3.2.3. Wheat Germ CECF Translation Reaction

1. Prepare the feeding solution for translation reaction by combining on ice the following components (*see* **Note 7**).

| | |
|---|---|
| $H_2O$ | 362 µL |
| Translation pre-MIX | 25 µL |
| 100 m$M$ Mg(OAc)$_2$ | 15 µL |
| 2.5 $M$ KOAc | 14 µL |
| 40% Glycerol | 25 µL |
| 3 m$M$ 19 Aa's (-leu) | 33 µL |
| 20 m$M$ Leucine | 5 µL |
| 3% NaN$_3$ | 5 µL |
| 500 m$M$ CrP | 16 µL |

Prepare 500 µL of feeding solution per each CECF reaction (*see* **Note 8**). Place solution in thermostat at 25°C.

**Table 2**
**Composition of the Reaction Mixture of the Wheat Germ CECF Translation System**

| Component | Final concentration | Component | Final concentration |
|---|---|---|---|
| WGE | [30% (v/v)] | GTP | 0.4 mM |
| mRNA | [0.2 μM] | ATP | 1.0 mM |
| tRNA yeast | 0.05 mg/mL | CrP | 16 mM |
| CPK | 0.1 mg/mL | Amino acids | 0.2 mM* |
| HPRI | 0.5 U/μL | Spermidine | 0.25 mM |
| HEPES-KOH | 26.0 mM* | DTT | 1.6 mM* |
| Mg(OAc)$_2$ | [3.0 mM*] | Glycerol | 2% |
| KOAc | [70 mM*] | NaN$_3$ | 0.03% |

Concentrations marked with asterisk are calculated as a sum of increments added as individual component and as a part of WGE solution (e.g., 1.5 mM Mg[OAc]$_2$ is added as the salt solution, another 1.5 mM comes with 30% [v/v] of WGE containing 5 mM Mg[OAc]$_2$). Concentrations in brackets should be optimized for particular batch of WGE and template construct. Feeding solution contains only the components shown in gray area.

The mRNA concentration is given for the particular mObe-GFP-TMV construct. For other mRNAs, especially possessing different UTRs the optimal concentration may differ and should be determined for maximal yield (*see* **Note 7**).

2. Prepare the translation reaction mixture for translation reaction by combining on ice the following components.

| | |
|---|---|
| $H_2O$ | 23.7 µL |
| 20X Translation pre-MIX | 3.0 µL |
| 100 m$M$ Mg(OAc)$_2$ | 0.9 µL |
| 2.5 $M$ KOAc | 1.0 µL |
| 40% Glycerol | 3.0 µL |
| 3 m$M$ 19 Aa's (-leu) | 2.0 µL |
| 20 m$M$ Leucine | 0.3 µL |
| 3% NaN$_3$ | 0.6 µL |
| 500 m$M$ CrP | 1.9 µL |
| 2 mg/mL Yeast tRNA | 1.5 µL |
| 10 mg/mL CPK | 0.6 µL |
| 40 U/µL HPRI | 0.8 µL |
| Wheat germ extract | 18.0 µL |
| 5 µ$M$ mRNA mObe-GFP-TMV | 2.5 µL |

Prepare 60 µL per each CECF reaction (*see* **Subheading 2.2.1.**).
Final concentrations of the components in the translation reaction are shown in **Table 2**.

3. Transfer 55 µL of reaction mixture into the tube cap, cover the cap with membrane circle (use thin forceps; avoid trapping air bubble under the membrane). Center the membrane on the cap.

4. Close the cap with a tube body: position the tube vertically over the cap and press it straight down firmly without displacing the membrane.

5. Fill the reservoir (tube) with 0.5 mL of feeding solution, insert the magnetic bar and seal the opening with Parafilm. Use another Parafilm strip to seal the contact between a cap and a tube to prevent drying through the membrane edge.

6. Incubate the reactor for 24 h on a magnetic stirrer (120 rpm) in a thermostat at 25°C (*see* **Note 8**).

7. After the incubation is finished, remove the feeding solution thoroughly, take out the magnetic bar. Pierce and cut the membrane with pipet tip and collect the reaction mix.

8. Centrifuge the reaction mixture for 10 min at 14000$g$. Collect the supernatant and transfer it to the fresh tube. Reconstitute the pellet in the same volume of WGE buffer or Tris-buffered 8 $M$ urea.

9. Analyze the concentration of the product in the supernatant (soluble product) and in the pellet (insoluble product) by SDS-PAGE or by a suitable product activity test (soluble product).

### 3.2.4. Wheat Germ CECF Transcription-Translation Reaction

CECF transcription-translation reaction with wheat germ extract is performed analogous to the translation reaction. Follow the **steps 1–9** in **Subheading 3.2.3.** but prepare the feeding solution and the reaction mixture as given in the next paragraph. Plasmid or PCR fragment and T7 RNA polymerase are added instead of mRNA. Magnesium concentration in reaction mixture is modified and differs from that in feeding solution (*see* **Note 9**). Transcription-translation pre-Mix is used, containing all four NTPs and higher spermidine concentration.

Preparation of transcription-translation feeding solution, per total volume 500 μL.

| | |
|---|---|
| $H_2O$ | 362 μL |
| Transcription-translation pre-MIX | 25 μL |
| 100 m$M$ Mg(OAc)$_2$ | 15 μL |
| 2.5 $M$ KOAc | 14 μL |
| 40% Glycerol | 25 μL |
| 3 m$M$ 19 Aa's (-leu) | 33 μL |
| 20 m$M$ Leucine | 5 μL |
| 3% NaN$_3$ | 5 μL |
| 500 m$M$ CrP | 16 μL |

Preparation of transcription-translation reaction mixture, per total volume 60 μL.

| | |
|---|---|
| $H_2O$ | 22.4 μL |
| Transcription-translation pre-MIX | 3.0 μL |
| 100 m$M$ Mg(OAc)$_2$ | 1.2 μL |
| 2.5 $M$ KOAc | 1.0 μL |
| 40% Glycerol | 3.0 μL |
| 3 m$M$ 19 Aa's (-leu) | 2.0 μL |
| 20 m$M$ Leucine | 0.3 μL |
| 3% NaN$_3$ | 0.6 μL |
| 500 m$M$ CrP | 1.9 μL |
| 2 mg/mL tRNA yeast | 1.5 μL |
| 10 mg/mL CPK | 0.6 μL |
| 40 U/μL RNasin | 0.8 μL |
| 100 U/μL T7 RNA polymerase | 1.0 μL |
| Wheat germ extract | 18.0 μL |
| 2 mg/mL pObe-GFP-STNV | 2.5 μL |

Final concentrations of the components in the transcription-translation reaction are shown in **Table 3**.

# Table 3
## Composition of the Wheat Germ CECF Transcription-Translation Reaction (*see* Note 9)

| Component | Final concentration | Component | Final concentration |
|---|---|---|---|
| WGE | [30% (v/v)] | ATP | 1.0 m$M$ |
| Plasmid | [100 μg/mL] | GTP | 0.4 m$M$ |
| T7 RNA pol | [4 U/μL] | CTP | 0.4 m$M$ |
| tRNA yeast | 0.05 mg/mL | UTP | 0.4 m$M$ |
| CPK | 0.1 mg/mL | CrP | 16 m$M$ |
| HPRI | 0.5 U/μL | Amino acids | 0.2 m$M$* |
|  |  | Spermidine | 0.45 m$M$ |
| HEPES-KOH | 26.0 m$M$* | DTT | 1.6 m$M$* |
| Mg(OAc)$_2$ | [3.0 m$M$*] | Glycerol | 2% |
| KOAc | [70 m$M$*] | NaN$_3$ | 0.03% |

Concentrations of the components marked with asterisk are calculated as a sum of increments added as individual component and as a part of WGE solution (e.g., 1.5 m$M$ Mg[OAc]$_2$ is added as the salt solution, another 1.5 m$M$ comes with 30% [v/v] of WGE containing 5 m$M$ Mg[OAc]$_2$). Note that magnesium concentration in the combined transcription-translation reaction is set higher in the reaction mixture (3.5 m$M$) to promote initial accumulation of the transcript. In less than 1 h the magnesium concentration in reaction mixture is equilibrated with that in feeding solution.

Components in gray area are added to the reaction mixture only. Values in brackets are subjected to change with different WGE or templates and should be optimized for each set.

## 4. Notes

1. Access to the reaction mixture is necessary when the wheat germ translation system is set to run for more then 24 h. In that case readdition of mRNA to the reaction mixture is performed every 24 h. Also, if the synthesis time course is of interest, the samples should be taken at different time points from the reaction mixture to analyze the product concentration (*see* also **Note 12**).

2. Other strains of *E. coli* can be used (e.g. BL21[DE3]). The mutant *E. coli* strains with reduced degradative activities were used for increasing protein yield in a cell-free system *(22)*.

3. In many *E. coli* expression vectors the ribosome-binding site (RBS) of gene 10 of T7 phage is used. This sequence provides efficient expression in the *E. coli* cell-free system as well. Expression in wheat germ system requires special 5'- and 3'-UTR sequences flanking the coding sequence (for details *see* **Note 14**). Presence of affinity tag sequence (e.g., His$_6$ tag) at 5'- or 3'-end of coding sequence serves for one step purification of the product. In the commercial RTS expression systems (Roche Diagnostics GmbH) a special pIVEX vector series are used optimized for in vitro expression in either *E. coli* or wheat germ systems and product purification.

4. Cell growth must be stopped by fast cooling of the cell culture to at least 10°C before collecting the cells. This can be performed by pouring the cell culture onto a crashed frozen 0.14 *M* NaCl or by cooling in a fermenter with high capacity cooling device, if available. Note that cell culture grown in flasks (overnight culture) is not suitable for S-30 extract preparation.

5. Quality of the extract strongly depends on the source of embryos. One should select the sort and batch of wheat seeds, which give embryos with minimal amount of foreign material stuck to the embryo. The selection should be finally made by testing the activity and duration of the cell-free translation system based on the prepared extract. Wheat germ preparations, which can be obtained from mill factories are often the crushed embryos contaminated with a large portion of endosperm. These crushed embryos cannot be used in this method.

6. The final inspection of embryos preparation is an important step. The material obtained by flotation still contains some fragments of endosperm and seed envelope stuck together or to the embryos. These fragments must be eliminated, and only undamaged and clean embryos must be used for extract preparation.

7. The optimal concentrations of $Mg^{++}$ and $K^+$ may vary for different wheat germ extract preparations in the range of 1.5 to 3.5 m$M$ and 60 to 120 m$M$, respectively, and have to be determined by performing a set of batch reactions at different salt concentrations. The optimal mRNA concentration may also vary for different mRNA constructs with different UTRs and coding sequences (*see* **Notes 13** and **14**). It also should be determined by performing a set of reactions in batch format. Increasing the concentration of WGE up to 50 vol percentage may be useful, this depends on the preparation of WGE and should be tested in batch reactions.

8. If the feeding solution/reaction mixture volume ratio is less than 10, the feeding solution has to be replaced after 8 h of the run. In some extracts uncoupled substrate degradation may be high. Replacement of the feeding solution after 8 h is helpful in this case irrespectively of volumes ratio to support the steady synthesis rate. Prepare twofold volume of feeding solution in **step 4** and keep excess volume at 4°C (do not freeze). Warm up the feeding solution to 25°C before replacement. For the prolong run (several days) prepare two portions of feeding solution each day and replace it every 12 h (*see* **Notes 10** and **11**).

9. Magnesium concentration in the combined transcription-translation reaction is set higher at the beginning of the reaction to promote initial accumulation of the transcript. In less than 1 h the magnesium concentration in reaction mixture is equilibrated with that in feeding solution. Plasmid concentration is given for the particular pObe-GFP-STNV construct. For other plasmid or PCR templates, especially those encoding different UTRs the optimal concentration may differ and should be determined for maximal yield (*see* **Note 7**).

10. During the protein synthesis reaction two types of substrates are being used: amino acids as monomers for synthesis of polypeptide chains, and nucleoside triphosphates as energy supply for translation factors and aminoacyl-tRNA synthetases and as monomers for synthesis of mRNA in transcription-translation systems. In the CECF systems the amino acids and energy substrates are constantly supplied and the products including $P_i$ are removed by diffusion exchange, and therefore the steady-state concentrations of the substrates and products are maintained during the run. In the case of the CECF systems, the feeding solution can be replaced in certain time intervals in order to maintain the initial energy compound concentration outside the reaction compartment. The replacement of feeding solution (and addition of mRNA) is necessary in wheat germ system to make it running for several days. Nevertheless, the problem still exists if the rate of exchange is not sufficient to compensate both the coupled and uncoupled consumption of substrates in the reaction mixture.

It was found that in bacterial cell-free system some amino acids are actively metabolized and their concentrations decrease rapidly *(23–26)*. To oppose this effect six amino acids (R, C, W, M, D, E) were added in higher concentration than others in the described *E. coli* CECF system. The cell extracts from mutant *E. coli* strains with reduced degradative activities were also successfully used for cell-free system *(22)*.

11. Because of NTPase and phosphatase activities in cell extracts, rapid uncoupled hydrolysis of NTPs occurs in the incubation mixture, in addition to their productive consumption during protein synthesis. To support the energy potential of the translation system the high-energy phosphate donors are used for regeneration of nucleoside triphosphates. In a wheat germ system creatine phosphate is the only energy substrate currently used. In bacterial system phosphoenol pyruvate is commonly used, acetyl phosphate *(27)* and creatine phosphate *(11)* are other options. According to our experience the combination of PEP and AcP described in the present protocol ensured higher activity of *E. coli* system, than PEP or AcP alone. Recently several novel NTP regeneration systems has been proposed, which increase the yield of protein synthesis in batch version of bacterial cell-free system *(28–31*; and this volume, p. 3). These regeneration systems use the oxidation of substrates of glycolytic pathway (pyruvate or glucose-6-phosphate) to regenerate the ATP and thus avoid the accumulation of inorganic phosphate, which is inhibitory to the translation system. These new regeneration systems can be useful for the continuous cell-free systems as well.

12. Although the dialysis membrane reactor is the simplest implementation of the CECF principle, several other approaches have been used successfully. In a hollow fiber reactor a bundle of hollow fibers substituted for the dialysis membrane *(32,33)* . In column reactor constructions the reaction mixture was encapsulated in the semipermeable vesicles surrounded by flow of feeding solution *(32)*, or the reaction mixture was passed through the Sephadex column equilibrated in feeding solution *(34)*. Another approach was suggested for micro scale high-throughput screening expression: the exchange proceeds across the liquid boundary between two layers, the reaction mixture presents the lower layer and the feeding solution is the upper one *(35)*.

13. Maintenance of the appropriate mRNA concentration during the long runtime is another key point of the continuous cell-free expression systems. RNase activities present in cell extracts usually restrict the lifetime of mRNA. Combined transcription-translation systems where mRNA is continuously renewed by ongoing transcription of DNA template has

practically no alternative for the continuous systems based on crude bacterial extracts rich in RNase activities. Circular plasmid DNA is commonly used as template for transcription in the prokaryotic system. The presence of the classic Shine-Dalgarno sequence and the epsilon enhancer of T7 phage gene10 upstream the initiation codon are favorable for efficient initiation of translation in the prokaryotic cell-free systems. A stable hairpin at the 3'-end of mRNA contributes to the protection of mRNA against 3'-exonucleolytic degradation. A significant stability of mRNA can be achieved by insertion of the coding sequence between mutually complementary 5'- and 3'-terminal untranslated sequences forming a stable stem helix (36). In the wheat germ system with lower nuclease activities both plasmid DNAs and PCR fragments serve well. As the excess of mRNA may inhibit translation in eukaryotic systems, the rate of transcription must be adjusted to the level that prevents overproducing mRNA. In the CECF systems the transcription rate can be easily regulated during run by changing the $Mg^{2+}$ and NTP concentrations via dialysis *(37)*. It is advantageous to start the combined transcription-translation run with a higher $Mg^{2+}$ concentration when transcription is of high rate and mRNA accumulates, and then continue with a lower $Mg^{2+}$ concentration that provides a modest level of mRNA synthesis compensating the degradation. An example of GFP synthesis in CECF system using this approach is given in **Fig. 3**. Another way is the periodical readdition of mRNA into the reaction mixture; this is practical in a wheat germ translation system. Addition of mRNA every 24 h provides the wheat germ translation system running with a constant rate for several days *(12)*.

14. In the eukaryotic systems a useful element for efficient initiation of translation is the so-called Kozak consensus sequence around initiation AUG codon *(38)*. The eukaryotic mRNAs are usually capped at the 5'-ends and polyadenylated at the 3'-ends in order to be well expressed. The 5'-cap structure and the 3'-poly(A) tail increase both the translation initiation efficiency and the stability of eukaryotic mRNAs. In the cell-free systems where translation is combined with transcription, the mRNA transcripts appear without caps. The addition of enzymatic systems for capping and polyadenylation into cell extracts is inconvenient and expensive. At the same time, RNAs of some plant viruses can effectively initiate translation without caps. Introducing the 5'- and 3'-elements of viral RNAs in mRNA constructs allow to circumvent the problems with capping and polyadenylation of eukaryotic mRNAs in cell-free translation. RNA of many plant viruses with RNA genomes, such as tobacco mosaic virus (TMV), alfalfa mosaic virus (AMV), brome mosaic virus (BMV), turnip

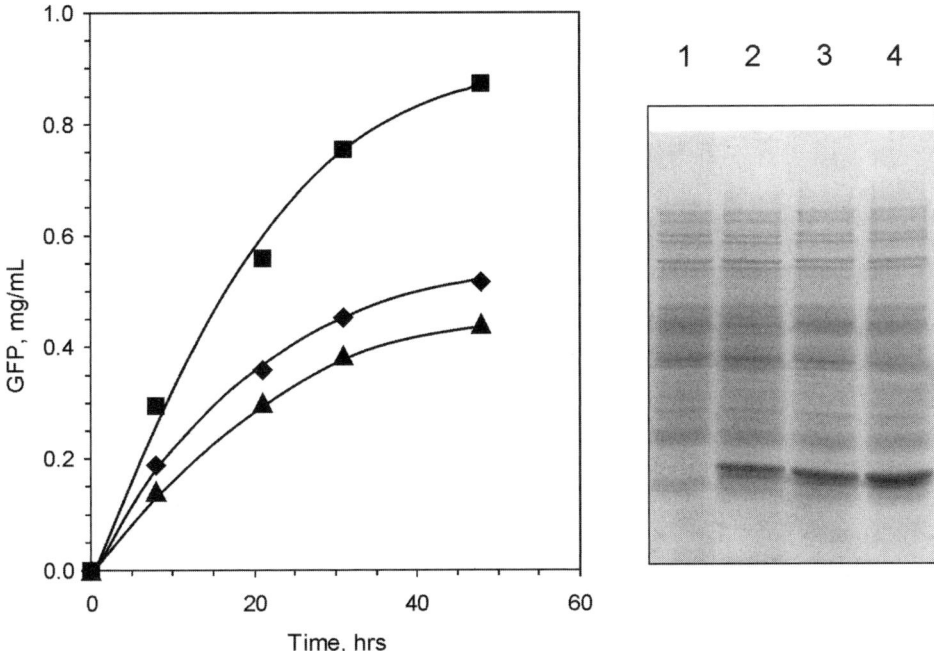

Fig. 3. Synthesis of GFP in a wheat germ continuous-exchange cell-free (CECF) transcription-translation system with different DNA templates, containing GFP gene with obelin 5'-UTR and STNV 3'-UTR under the control of T7 promoter *(48)*. (A) Time course of the protein synthesis with 0.02 mg/mL PCR fragment (▲), or 0.1 mg/mL plasmid DNA (◆), or 0.2 mg/mL plasmid DNA (■). Reactions were performed as described in Methods. GFP concentration was monitored by in situ fluorescence measurement in the cap microreactor during the CECF run and by direct fluorescence measurements against GFP standard after completing the run. B: 12% SDS-PAGE analysis of the corresponding reaction mixtures after 48 h CECF run (lanes 2–4, lane 1, reaction mixture at 0 h). Samples were depleted of ribosomes by centrifugation over the sucrose cushion for 15 min in TLA 100 rotor. Two microliters of the reaction mixture were loaded in each lane. Staining with Coomassie G 250.

yellow mosaic virus (TYMV), are not polyadenylated and, instead, have pseudoknot/tRNA-like 3'-UTRs (reviewed in **ref. *39***). The 5'-untranslated regions (5'-UTRs) of satellite tobacco necrosis virus (STNV) RNA and barley yellow dwarf virus (BYDV-PAV) RNA can serve as powerful translation enhancers *(40–42)*, whereas their long 3'-UTR sequences may contribute to the stability of mRNA. It has been demonstrated that these 3'-untranslated regions (3'-UTRs) can effectively replace the poly(A)-tail in translation when attached to heterologous (nonviral) coding sequences both in plant and animal systems, in vivo and in vitro *(43–46)*. It was reported that the original 143 nt 5'-UTR of TEV RNA can be trimmed to 29 nt and the truncated leader sequence still can be used for effective cap-independent translation of heterologous coding sequences in a wheat germ cell-free translation system *(47)*. Complementarity of 5'-UTR to the specific region of 18S rRNA was shown to provide an efficient translation initiation in the wheat germ system *(48)*. Recently we have found that a nonviral 5'-UTR of obelin mRNA is an efficient leader for expression of foreign coding sequences in combination with 3'-UTRs of viral origin in wheat germ CECF system *(49)*. The important feature of practical importance is that no inhibition with excess mRNA possessing obelin 5-UTR is observed in wheat germ cell-free system. Translation of different mRNA constructs with this leader in wheat germ CECF system is shown in **Fig. 4** (K. V. Gromova, V. A. Shirokov, unpublished results).

15. How to rise correct protein folding in cell-free systems. In some cases a significant portion of a polypeptide synthesized in a cell-free system appears in a misfolded form. Such a "nonnative" (denatured) protein is functionally inactive, tends to form aggregates and often precipitates. At the same time, the proportion of the misfolded protein in the product of a cell-free translation system strongly depends on the physical and chemical conditions during incubation. In most cases the optimization of the process of cotranslational folding in cell-free translation systems can result in the synthesis of correctly folded, functionally active proteins with the yield up to 90–100% of the correctly folded molecules in the total polypeptide product. Among the parameters to be optimized are the following: (1) the *temperature* of the reaction mixture during translation (not necessarily optimal for the synthesis rate); (2) the *concentration of cell-free extract* used for translation (not necessarily the higher the better); (3) the presence in the incubation mixture during translation of *low-molecular-weight ligands*, such as cofactors, prosthetic groups, substrates (or their analogs) of the protein synthesized; (4) the *redox potential* of the incubation mixture and the presence of *protein disulfide isomerases, molecular chaperones* and other protein folding catalysts, when necessary, and especially in the case of the synthesis of disulfide-bonded proteins; (5) the addition of a *nonionic detergent* to the incubation mixture.

In principle, the successful synthesis of functionally active proteins of different origin can be achieved in cell-free systems based on both bacterial and eukaryotic extracts (*see*, e.g., **ref. 50**). Unsuccessful attempts to syn-

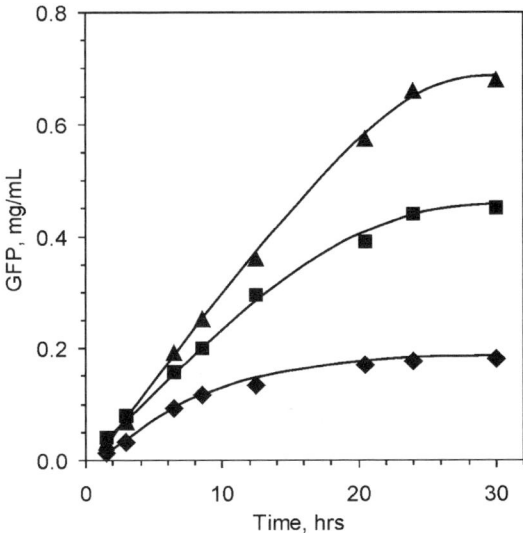

Fig. 4. Synthesis of GFP in a wheat germ continuous-exchange cell-free (CECF) translation system on mRNA with different 5'- and 3'-UTRs. Reactions were performed as described in **Subheading 3**. Reaction mixtures contained the following mRNA in 500 µM concentration: 5'-UTR$_{obelin}$-*GFP*-3'-UTR$_{TMV}$ (▲), 5'-UTR$_{TMV}$-*GFP*-3'-UTR$_{TMV}$ (■), or 5'-UTR$_{obelin}$-*GFP*-3'-UTR$_{STNV}$ (◆). GFP concentration was monitored by in situ fluorescence measurement in the cap microreactor during the CECF run and by direct fluorescence measurements against GFP standard after completing the run.

thesize some mammalian proteins in active form using a bacterial extract seem to be the result of nonoptimized folding conditions, particularly too fast elongation rate on bacterial ribosomes; an artificial retardation of elongation may help in this case.

a. The temperature of reaction mixture for protein synthesis is one of the most important parameter determining the correct folding of newly synthesized polypeptides. The temperature optimal for cotranslational protein folding in cell-free systems is not necessarily the optimum for the synthesis rate. For instance, the synthesis rate in *E. coli* systems is the highest at 37°C or slightly above, but in most cases the yield of a *functionally active protein* is higher at lower temperatures. An example is presented in **Fig. 5**: it is seen that the specific activity of firefly (*Photinus pyralis*) luciferase synthesized in the *E. coli* translation system is the highest when synthesized at 30°C and three times lower when synthesized at 25 or 35°C.

Another example is the synthesis of human proinsulin in *E. coli* translation system: in this case it was found that decreasing the temperature of incubation to 25°C resulted in significant improvement of proinsulin solubility (up to 95%), whereas the total yield of the polypeptide diminished just slightly *(51)*. Thus, it seems that, in order to synthesize a functionally active protein with the maximal specific activity, the temperature during protein synthesis in cell-free translation systems must be optimized in each case depending on a protein to be produced.

Two plausible explanations of temperature optima for correct protein folding can be considered. (1) In the case of cotranslational protein folding, the elongation of a growing polypeptide at a higher temperature can be too fast, and the sections of the chain have no time for completion of correct local folds and thus interfere with following sections. (At a lower temperature the growing polypeptide may be too rigid for adequate conformational search.) (2) Irrespective of elongation rate, the fast search for correct conformations requires annealing ("subdenaturing") conditions where a correct conformation is still stable but intermediate and incorrect conformations are unstable and transitory; it is the "subdenaturing" temperature that may be optimal for correct protein folding during translation.

b.  It is often believed that the higher is the concentration of a cell extract (that includes translation factors and other components of protein-synthesizing machinery) in a cell-free system, the better is the yield of a protein to be synthesized. Indeed, in a number of cases concentrated cell extracts were successfully used in effective cell-free translation systems *(11,22,52,53)*. However, this is not a common rule. For instance, in the case of human proinsulin synthesis the proportion of the bacterial S30 extract in the transcription-translation reaction was found to be critical for synthesis of the protein in a soluble form and the lower concentrations were more advantageous (**Table 4**). The optimal concentration of the extract providing both solubility and high yield of proinsulin ranged around 18% *(51)*. As it is seen in the table, the elevation of cell extract concentration resulted neither in the improvement of protein folding nor in a significant rise of the protein yield. Thus, highly concentrated extracts are not always beneficial for the synthesis of correctly folded proteins in cell-free systems. Generally, it can be recommended to optimize the proportion of a cell extract in a cell-free incubation mixture for each type and lot of the extract and, maybe, for each protein to be synthesized.

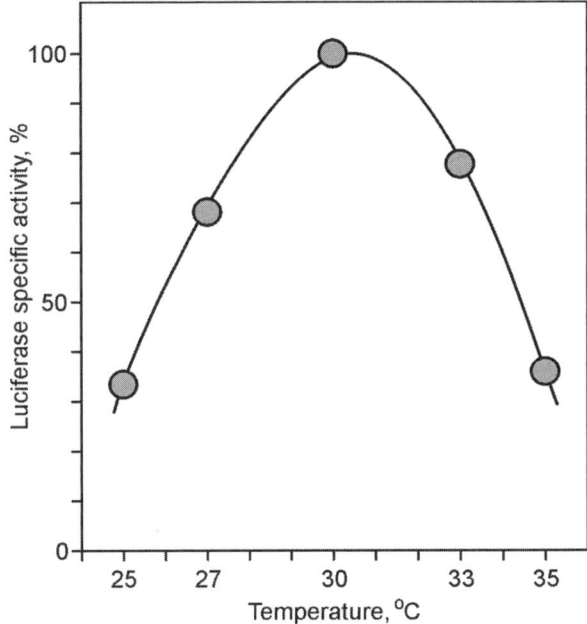

Fig. 5. Specific activity of luciferase as a function of temperature in a cell-free translation system. Translation in 30% *Escherichia coli* S30 extract was carried out at indicated temperatures in the presence of 1.2 m$M$ ATP, 0.8 m$M$ GTP, 0.03 mg/mL folinic acid, 80 m$M$ creatine phosphate, 0.25 mg/mL creatine kinase, 4% PEG 8000, 0.34 m$M$ amino acids each except leucine, 4.0 μ$M$ [$^{14}$C]leucine, and 0.175 mg/mL total *E. coli* tRNA in a buffer containing 10 m$M$ Mg(OAc)$_2$, 26 m$M$ HEPES-KOH pH$_{20}$ 7.6, 100 m$M$ KOAc, and 1.7 m$M$ dithiotreitol. The reaction mixture also contained 0.1 m$M$ luciferin. Translation was initiated by addition of mRNA (transcribed from the plasmid pT7 luc; **ref. 50**) to the concentration of 74 n$M$. To determine enzymatic activity of luciferase, 10-mL aliquot was withdrawn from the translation mixture after incubation at given temperature and immediately added to 10 mL of a solution containing 40 μ$M$ thiostrepton, 1.2 m$M$ ATP, 0.1 m$M$ luciferin, 10 m$M$ Mg(OAc)$_2$, 26 m$M$ HEPES-KOH pH$_{20}$ 7.6, 100 m$M$ KOAc, 1.7 m$M$ dithiothreitol; the intensity of emitted light in the aliquots was recorded at 25°C in a luminometer. The same aliquots were then analyzed with 10% SDS-PAGE, and the radioactivity of the full-length luciferase bands was determined. Specific activity was estimated as the ratio of the luminescence to the amount of the synthesized protein calculated from radioactivity. (T. Kazakov and V. Kolb, unpublished.)

c. Specific ligands, such as cofactors, prosthetic groups, substrates and their analogs, can bind to their proteins before the completion of their synthesis during translation, i.e., cotranslationally (*see*, e.g., **refs. 54 and 55**). From this it is not surprising that the synthesis of a protein in the presence of its specific ligand can improve folding of the protein *(56–58)*. For example, the presence of flavin adenine dinucleotide (FAD) during cell-free synthesis of thioredoxin reductase resulted in fourfold rise of the specific activity of the enzyme *(57)*. In the case of firefly luciferase synthesized in the presence of its substrate luciferin, the specific activity of the enzyme was high, irrespective of whether this protein was synthesized in eukaryotic or prokaryotic translation systems *(50)*. At the same time the synthesis of the enzyme in the absence of luciferin produced the protein with low specific activity, which could be improved by the addition of trigger factor during translation *(59)*. On the whole, the presence of specific ligands during cell-free synthesis is highly promotional for correct folding of synthesized proteins.

d. Many proteins can be synthesized and correctly folded in cell-free systems without molecular chaperones (*see*, e.g., **ref. 60**). In this laboratory the contribution of molecular chaperones of the Hsp70 family to cotranslational folding of firefly luciferase was assessed in *E. coli* cell-free translation system preliminarily deprived of the chaperones. It was found that the specific activity of the luciferase synthesized was high in the absence of chaperones and the addition of chaperones did not increase it further (**Fig. 6**; *see* **ref. 60a**).

In some cases, however, misfolding and precipitation of the product are observed. To improve the situation the addition of molecular chaperones to cell-free translation systems was often practiced (*see*, e.g., **refs. 61** and **62**, and papers in **refs. 63** and **64**). The best effects were mentioned with the synthesis of disulfide-bonded proteins when combinations of chaperones and a protein disulfide isomerase were used, such as in the syntheses of manganese peroxidase *(58)*, single-chain antibodies and Fab fragments *(61,65)*, soluble tissue plasminogen activator *(66)*, granulocyte-macrophage colony-stimulating factor *(67)* and protease domain of murine urokinase *(68)*. The activities of some other proteins (e.g., mitochondrial rhodanese **ref. 62** and the catalytic subunit of human telomerase **ref. 69**) sometimes were also strongly dependent on the presence of molecular chaperones in a cell-free system, but in most cases no or just marginal effects of chaperone addition were observed for proteins without disulfide bonds.

**Table 4**
**Dependence of the Solubility of Newly Synthesized Proinsulin on the Concentration of *E. coli* S30 Extract in a Cell-Free Transcription-Translation System[a]**

| Concentration of *E. coli* S30 extract in incubation mixture | Soluble proinsulin, mg per milliliter of incubation mixture | Aggregated proinsulin, mg per milliliter of incubation mixture | Soluble proinsulin fraction, % |
|---|---|---|---|
| 8.8% | 0.66 | 0.03 | 96 |
| 17.5% | 2.7 | 0.20 | 93 |
| 35% | 2.96 | 1.30 | 69 |
| 52.5% | 3.32 | 3.78 | 47 |

[a]CECF transcription-translation was performed in the presence of [$^{14}$C]Leu at different S30 extract concentrations. Human proinsulin containing C-terminal hexa-histidine tag was synthesized from the plasmid pETInsHis$_6$ (*51*). After 24 h reaction, mixtures were fractionated by centrifugation at 12000g for 15 min at 20°C. Aliquots of pellet and supernatant fractions were analysed by SDS PAAGE and radioactivity was determined in proinsulin bands.

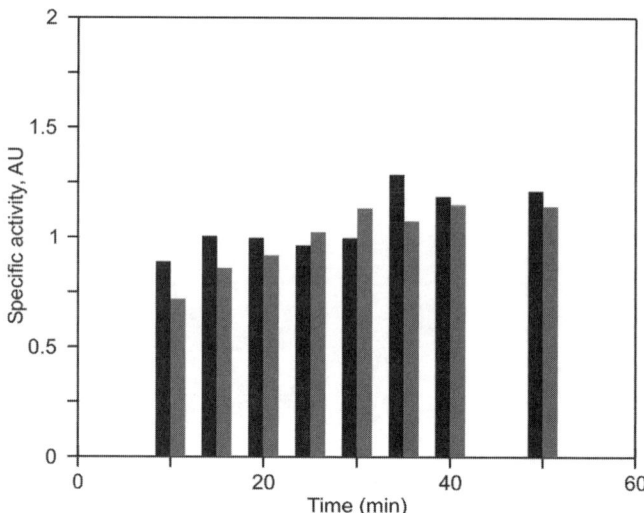

Fig. 6. Specific activity of luciferase synthesized in the presence of added Hsp70 chaperones. Translation and specific activity determination were performed in *Escherichia coli* S30 translation system as described in the legend to **Fig. 5** except the presence of [$^{14}$C]phenylalanine instead of [$^{14}$C]leucine. Translation was performed at 25°C. Black symbols correspond to the reaction with addition of Hsp70 chaperones, gray symbols – the reaction without added chaperones. Individual DnaK, DnaJ, and GrpE (dimer) were used in concentrations up to 1.3 µ*M*, 200 µ*M*, and 650 µ*M*, respectively. (*See* **ref. 60a**)

e.  The introduction of nonionic detergents into a cell-free translation system can be very helpful in preventing product aggregation and improving the solubility and specific activity of the proteins synthesized (*see*, e.g., **refs. 70–74** and this volume, p. 57 and other papers in **refs. 63** and **64**). In this laboratory the following detergents were successfully tested in the wheat germ cell-free translation system programmed with DHFR (dihydrofolate reductase) mRNA: Brij 35, digitonin, Triton X-100, Nonidet P40, and octyl glucoside. No decrease in the protein synthesizing activity was detected in the system with each of these detergents at the concentrations ranged from 0.2 to 1%, except octyl glucoside (the latter inhibited translation at 0.2% concentration and causes a complete cessation of the synthesis at 1%).

In the case of human proinsulin synthesis in the bacterial S30 system, the addition of 0.2% Brij 35 to the reaction mixture showed a significant increase in solubility of the synthesized protein without serious loss of the protein yield *(51)*. Digitonin added to the system to the concentration of 0.5% did not change the protein yield as well, although only 50% increase in solubility was observed.

## Acknowledgments

The authors thank Kira V. Gromova, Irina G. Dashkova, Maxim S. Svetlov, Tejmuraz S. Kazakov, and Gelina S. Kopeina for performing the experiments mentioned in the chapter. We are grateful to Lyubov A. Shaloiko for cooperation and helpful discussions, and Elena R. Arutyunyan and Svetlana N. Proshkina for excellent technical assistance. This work was supported by Roche Diagnostics GmbH, by grant no. 1966.2003.4 from the President of the Russian Federation, and by the Program on Molecular and Cellular Biology of the Russian Academy of Sciences.

## References

1. Spirin, A. S., Baranov, V. I., Ryabova, L. A., Ovodov, S. Y., and Alakhov, Y. B. (1988) A continuous cell-free translation system capable of producing polypeptides in high yield. *Science* **242,** 1162–1164.
2. Baranov, V. I., Morozov, I. Y., Ortlepp, S. A., and Spirin, A. S. (1989) Gene expression in a cell-free system on the preparative scale. *Gene* **84,** 463–466.
3. Baranov, V. I. and Spirin, A. S. (1993) Gene expression in cell-free systems on preparative scale. *Methods Enzymol.* **217,** 123–142.
4. Alakhov, Y. B., Baranov, V. I., Ovodov, S. Y., Ryabova, L. A., and Spirin, A. S. (1995) Method of preparing polypeptides in cell-free translation system. *United States Patent no.* 5,478,730.
5. Chekulayeva, M. N., Kurnasov, O. V., Shirokov, V. A., and Spirin, A. S. (2001) Continuous-exchange cell-free protein-synthesizing system: Synthesis of HIV-1 antigen Nef. *Biochem. Biophys. Res. Commun.* **280,** 914–917.
6. Martemyanov, K. A., Shirokov, V. A., Kurnasov, O. V., Gudkov, A. T., and Spirin, A. S. (2001) Cell-free production of biologically active polypeptides: Application to the synthesis of antibacterial peptide Cecropin. *Protein Expr. Purif.* **21,** 456–461.
7. Shirokov, V. A., Simonenko, P. N., Biryukov, S. A., and Spirin, A. S. (2002) Continuous-flow and continuous-exchange cell-free translation systems and reactors, in *Cell-Free Translation Systems,* (Spirin, A. S., ed.), Springer-Verlag, Berlin, Germany, pp. 91–107.
8. Spirin, A. S. (2004) High-throughput cell-free systems for synthesis of functionally active proteins. *Trends Biotech.* **22,** 538–545.

9. Kim, D. -M. and Choi, C. -Y. (1996) A semicontinuous prokaryotic coupled transcription/translation system using a dialysis membrane. *Biotechnol. Prog.* **12,** 645–649.

10. Davis, J., Thompson, D., and Beckler, G. S. (1996) Large scale dialysis cell-free system. *Promega Notes Magazine* **56,** 14–21.

11. Kigawa, T., Yabuki, T., Yoshida, Y., et al. (1999) Cell-free production and stable-isotope labelling of milligram quantities of proteins. *FEBS Lett.* **442,** 15–19.

12. Madin, K., Sawasaki, T., Ogasawara, T., and Endo, Y. (2000) A highly efficient and robust cell-free protein synthesis system prepared from wheat embryos: plants apparently contain a suicide system directed at ribosomes. *Proc. Natl. Acad. Sci. USA* **97,** 559–564.

13. Martin, G. A., Kawaguchi, R., Lam, Y., DeGiovanni, A., Fukushima, M., and Mutter, W. (2001) High-yield, in vitro protein expression using a continuous-exchange, coupled transcription/translation system. *BioTechniques* **31,** 948–953.

14. Sawasaki, T., Ogasawara, T., Morishita, R., and Endo, Y. (2002) A cell-free protein synthesis system for high-throughput proteomics. *Proc. Natl. Acad. Sci. USA* **99,** 14,652–14,657.

15. Endo, Y. and Sawasaki, T. (2004) High-throughput, genome scale protein production method based on the wheat germ cell-free expression system. *J. Struct. Funct. Genomics* **5,** 45–57.

16. Yokoyama, S. (2003) Protein expression systems for structural genomics and proteomics. *Curr. Op. Chem. Biol.* 7, 39–43.

17. Betton, J. -M. (2003) Rapid translation system (RTS): a promising alternative for recombinant protein production. *Current Protein and Peptide Science* **4,** 73–80.

18. Sambrook, J., Fritsch, E. F., and Maniatis, T. (1989) *Molecular Cloning: a Laboratory Manual.* Cold Spring Harbor Laboratory, Cold Spring Harbor, NY.

19. Pokrovskaya, I. D. and Gurevich, V. V. (1994) *In vitro* transcription: preparative RNA yields in analytical reactions. *Anal. Biochem.* **220,** 420–423.

20. Zubay, G. (1973) In vitro synthesis of protein in microbial systems. *Annu. Rev. Genet.* **7,** 267–287.

21. Erickson, A. H. and Blobel, G. (1983) Cell-free translation of messenger RNA in a wheat germ system. *Methods Enzymol.* **96,** 38–50.

22. Michel-Reydellet, N., Calhoun, K., and Swartz, J. (2004) Amino acid stabilization for cell-free protein synthesis by modification of the *Escherichia coli* genome. *Metab Eng.* **6,** 197–203.

23. Kim, D. M., Kigawa, T., Choi, C. Y., and Yokoyama, S. (1996) A highly efficient cell-free protein synthesis system from *Escherichia coli*. *Eur. J. Biochem.* **239,** 881–886.

24. Patnaik, R. and Swartz, J. R. (1998) *E. coli*-based in vitro transcription/ translation: in vivo-specific synthesis rates and high yields in a batch system. *BioTechniques* **24,** 862–868

25. Kim, D. M. and Swartz, J. R. (2000) Prolonging cell-free protein synthesis by selective reagent additions. *Biotech. Prog.* **16,** 385–390

26. Kim, R. G. and Choi, C. Y. (2001) Expression-independent consumption of substrates in cell-free expression system from *Escherichia coli*. *J. Biotechnol.* **84,** 27–32

27. Ryabova, L. A., Vinokurov, L. M., Shekhovtsova, E. A., Alakhov, Y. B., and Spirin, A. S. (1995) Acetyl phosphate as an energy source for bacterial cell-free translation system. *Anal. Biochem.* **226,** 184–186

28. Kim, D. M. and Swartz, J. R. (1999) Prolonging cell-free protein synthesis with a novel ATP regeneration system. *Biotech. Bioengineering* **66,** 180–188

29. Kim, D. -M. and Swartz, J. R. (2001) Regeneration of adenosine triphosphate from glycolytic intermediates for cell-free protein synthesis. *Biotech. Bioeng.* **74,** 309–316.

30. Jewett, M. C. and Swartz, J. R. (2004) Mimicking the *Escherichia coli* cytoplasmic environment activates long-lived and efficient cell-free protein synthesis. *Biotech. Bioeng.* **86,** 19–26.

31. Jewett, M. C. and Swartz, J. R. (2004) Rapid expression and purification of 100 nmol quantities of active protein using cell-free protein synthesis. *Biotechnol. Prog.* **20,** 102–109.

32. Spirin, A. S. (1992) Cell-free protein synthesis bioreactor, in *Frontiers in Bioprocessing II*, (Todd, P., Sikdar, S. K., and Beer, M., ed.), American Chemical Society, Washington, DC, pp. 31–43.

33. Nakano, H., Shinbata, T., Okumura, R., Sekiguchi, S., Fujishiro, M., and Yamane, T. (1999) Efficient coupled transcription/translation for PCR template by a hollow fiber membrane bioreactor. *Biotechnol. Bioeng.* **64,** 194–199.

34. Buchberger, B., Mutter, W., and Röder, A. (2002) Matrix reactor: a new scalable principle for cell-free protein expression, in *Cell-Free Translation Systems,* (Spirin, A. S., ed.), Springer, Germany, pp. 121–128.

35. Sawasaki, T., Hasegawa, Y., Tsuchimochi, M., et al. (2002) A bilayer cell-free protein synthesis system for high-throughout screening of gene products. *FEBS Lett.* **514,** 102–105.

36. Ugarov, V. I., Morozov, I. Yu., Jung, G. V., Chetverin, A. B., and Spirin, A. S. (1994) Expression and stability of recombinant RQ-mRNA in cell-free translation systems. *FEBS Lett.* **341,** 131–134.

37. Biryukov, S. V., Simonenko, P. N., Shirokov, V. A., and Spirin, A.S. (2004) Method for synthesis of polypeptides in cell-free systems. *United States Patent no.* 6,783,957.

38. Kozak, M. (1989) Context effects and inefficient initiation at the non-AUG codons in eucaryotic cell-free translation systems. *Mol. Cell. Biol.* **9,** 5073–5080.

39. Florentz, C. and Giegé, R. (1995) tRNA-like structures in plant viral RNAs, in *tRNA: Structure, Biosynthesis, and Function,* (Söll, D. and RajBhandary, U. L., eds.), ASM Press, Washington, pp. 141–163.

40. Danthinne, X., Seurinck, J., Meulewaeter, F., van Montagu, M., and Cornelissen, M. (1993) The 3' untranslated region of satellite tobacco necrosis virus RNA stimulates translation in vitro. *Mol. Cell. Biol.* **13,** 3340–3349.

41. Timmer, R. T., Benkowski, L. A., Schodin, D., et al. (1993) The 5' and 3' untranslated regions of satellite tobacco necrosis virus RNA affect translational efficiency and dependence on a 5' cap structure. *J. Biol. Chem.* **268,** 9504–9510.

42. Wang, S. and Miller, W. A. (1995) A sequence located 4.5 to 5 kilobases from the 5'-end of the barley yellow dwarf virus (PAV) genome strongly stimulates translation of uncapped mRNA. *J. Biol. Chem.* **270,** 13,446–13,452.

43. Gallie, D. R. and Walbot, V. (1990) RNA preudoknot domain of tobacco mosaic virus can functionally substitute for a poly(A) tail in plant and animal cells. *Genes Dev.* **4,** 1149–1157.

44. Gallie, D. R., Feder, J. N., Schimke, R. T., and Walbot, V. (1991) Functional analysis of the tobacco mosaic virus tRNA-like structure in cytoplasmic gene regulation. *Nucleic Acids Res.* **19,** 5031–5036.

45. Ryabova, L. A., Torgashov, A. F., Kurnasov, O. V., Bubunenko, M. G., and Spirin, A. S. (1993) The 3'-terminal region of alfalfa mosaic virus RNA4 facilitates the RNA entry into translation in a cell-free system. *FEBS Lett.* **326,** 264–266.

46. Zeyenko, V. V., Ryabova, L. A., Gallie, D. R., and Spirin, A. S. (1994) Enhancing effect of the 3'-untranslated region of tobacco mosaic virus RNA on protein synthesis in vitro. *FEBS Lett.* **354,** 271–273.

47. Kawarasaki, Y., Kasahara, S., Kodera, N., et al. (2000) A trimmed viral cap-independent translation enhancing sequence for rapid in vitro gene expression. *Biothechnol. Prog.* **16,** 517–521.

48. Akbergenov, R. Zh., Zhanybekova, S. Sh., Kryldakov, R. V., et al. (2004) ARC-1, a sequence element complementary to an internal 18S rRNA segment, enhances translation efficiency in plants when present in the leader or intercistronic region of mRNAs. *Nucleic Acids Res.* **32**, 239–247

49. Shaloiko, L. A., Granovsky, I. E., Ivashina, T. V., Ksenzenko, V. N., Shirokov, V. A., and Spirin, A. S. (2004) Effective non-viral leader for cap-independent translation in a eukaryotic cell-free system. *Biotechnol. Bioeng.* **88**, 730–739.

50. Kolb, V. A., Makeyev, E. V., and Spirin, A. S. (2000) Co-translational folding of an eukaryotic multidomain protein in a prokaryotic translation system. *J. Biol. Chem.* **275**, 16,597–16,601.

51. Kommer, A., Dashkova, I. G., Yesipov, R. S., Miroshnikov, A. I., and Spirin, A. S. (2005) Synthesis of functionally active human proinsulin in a cell-free translation system. *Dokl. Akad. Nauk.* **401**, 1-5 (Dokl. Biochem. Biophys. 401,154–158).

52. Nakano, H., Tanaka, T., Kawarasaki, Y., and Yamane, T. (1994) An increased rate of cell-free protein synthesis by condensing wheat-germ extract with ultrafiltration membranes. *Biosci. Biotechnol. Biochem.* **58**, 631–634.

53. Yamane, T., Kawarasaki, Y., and Nakano, H. (1995) In vitro protein biosynthesis using ribosome and foreign mRNA: an approach to construct a protein biosynthesizer. *Ann. NY Acad. Sci.* **750**, 146–157.

54. Komar, A. A., Kommer, A., Krasheninnikov, I. A., and Spirin, A. S. (1993) Cotranslational heme binding to nascent globin chains. *FEBS Lett.* **326**, 261–263.

55. Komar, A. A., Kommer, A., Krasheninnikov, I. A., and Spirin, A. S. (1997) Cotranslational folding of globin. *J. Biol. Chem.* **272**, 10,646–10,651.

56. Mouat, M. F. (2000) Dihydrofolate influences the activity of *Escherichia coli* dihydrofolate reductase synthesized de novo. *Int. J. Biochem. Cell Biol.* **32**, 327–337.

57. Knapp, K. G. and Swartz, J. R. (2004) Cell-free production of active *E. coli* thioredoxin reductase and glutathione reductase. *FEBS Lett.* **559**, 66–70.

58. Miyazaki-Imamura, C., Oohira, K., Kitagawa, R., Nakano, H., Yamane, T., and Takahashi, H. (2003) Improvement of $H_2O_2$ stability of manganese peroxidase by combinatorial mutagenesis and high throughput screening using in vitro expression with protein disulfide isomerase. *Protein Eng.* **16**, 423–428.

59. Agashe, V. R., Guha, S., Chang, H. -C., et al. (2004) Function of trigger factor and DnaK in multidomain protein folding: increase in yield at the expense of folding speed. *Cell* **117**, 199–209.

60. Shimizu, Y., Inoue, A., Tomary, Y., et al. (2001) Cell-free translation reconstituted with purified components. *Nat. Biotechnol.* **19,** 751–755.

60a. Svetlov, M. S., Kommer, A., Kolb, V. A., and Spirin, A. S. (2006) Effective cotranslational folding of firefly luciferase without chaperones of Hsp70 family. *Prot. Sci.* **15,** 242–247.

61. Ryabova, L. A., Desplancq, D., Spirin, A. S., and Pluckthun, A. (1997) Functional antibody production using cell-free translation: effects of protein disulfide isomerase and chaperones. *Nat. Biotechnol.* **15,** 79–84.

62. Kudlicki, W., Mouat, M., Walterscheid, J. P., Kramer, G., and Hardesty, B. (1994) Development of a chaperone-deficient system by fractionation of a prokaryotic coupled transcription/translation system. *Anal. Biochem.* **217,** 12–19.

63. Spirin, A. S., ed. (2002) *Cell-Free Translation Systems.* Springer-Verlag, Heidelberg, Berlin, New York.

64. Swartz, J. R., ed. (2003) *Cell-Free Protein Expression.* Springer-Verlag, Berlin, Heidelberg, New York.

65. Jiang, X., Ookubo, Y., Fujii, I., Nakano, H., and Yamane, T. (2002) Expression of Fab fragment of catalytic antibody 6D9 in an *Escherichia coli* in vitro coupled transcription/translation system. *FEBS Lett.* **514,** 290–294.

66. Yin, G. and Swartz, J. R. (2004) Enhancing multiple disulfide bonded protein folding in a cell-free system. *Biotechnol. Bioeng.* **86,** 188–195.

67. Yang, J., Kanter, G., Voloshin, A., Levy, R., and Swartz, J. R. (2004) Expression of active murine granulocyte-macrophage colony-stimulating factor in an *Escherichia coli* cell-free system. *Biotechnol. Prog.* **20,** 1689–1696.

68. Kim, D. -M. and Swartz, J. R. (2004) Efficient production of a bioactive, multiple disulfide-bonded protein using modified extracts of *Escherichia coli. Biotechnol. Bioeng.* **85,** 122–129.

69. Steigerwald, R., Nemetz, C., Walckhoff, B., and Emrich, T. (2002) Cell-free expression of a 127 kDa protein: the catalytic subunit of human telomerase, in *Cell-Free Translation Systems,* (Spirin, A. S., ed.), Springer-Verlag, Heidelberg, Berlin, New York, pp. 165–173.

70. Nemetz, C., Wessner, S., Krupka, S., Watzele, M., and Mutter, W. (2002) Cell-free expression of soluble human erythropoietin, in *Cell-Free Translation Systems,* (Spirin, A. S., ed.), Springer-Verlag, Heidelberg, Berlin, New York, pp. 157–163.

71. Maurer, P., Moratzky, A., Fecher-Trost, C., et al. (2003) Cell-free synthesis of membrane proteins on a preparative scale, in *Cell-Free Protein*

*Expression,* (Swartz, J. R., ed.), Springer-Verlag, Berlin, Heidelberg, New York, pp. 133–139.

72. Busso, D., Kim, R., and Kim, S. H. (2003) Expression of soluble recombinant proteins in a cell-free system using a 96-well format. *J. Biochem. Biophys. Methods* **55,** 233–240.

73. Busso, D., Kim, R., and Kim, S. H. (2004) Using an *Escherichia coli* cell-free extract to screen for soluble expression of recombinant proteins. *J. Struct. Funct. Genomics* **5,** 69–74.

74. Klammt, C., Löhr, F., Schafer, B., et al. (2004) High level cell-free expression and specific labeling of integral membrane proteins. *Eur. J. Biochem.* **271,** 568–580.

# 3

# Cell-Free Production of Integral Membrane Proteins on a Preparative Scale

Christian Klammt, Daniel Schwarz, Volker Dötsch, and Frank Bernhard

## Summary

The chapter will focus on the high level cell-free production of integral membrane proteins having multiple transmembrane segments by using an individual coupled transcription/translation system based on an *Escherichia coli* S30-extract. We describe in detail the setup and optimization of the cell-free expression technique to obtain the maximum yield of recombinant proteins. The protocol can be used for the expression of soluble membrane proteins as well as for their production as a precipitate. In addition, we will provide protocols for the efficient solubilization and reconstitution of membrane proteins directly from the cell-free produced precipitates.

**Key Words:** Integral membrane proteins; structural analysis; solubilization; reconstitution; detergent; NMR spectroscopy; amino acid transporter; G-protein coupled receptors; β-barrel proteins; multidrug resistance.

## 1. Introduction

Although membrane proteins (MPs) spanning lipid bilayers with multiple transmembrane segments (TMS) are highly abundant in fully

From: *Methods in Molecular Biology*, vol. 375,
*In Vitro Transcription and Translation Protocols, Second Edition*
Edited by: G. Grandi © Humana Press Inc., Totowa, NJ

sequenced genomes, only few high-resolution structures have been determined so far. This obvious discrepancy can mainly be attributed to the
tremendous difficulties that generally emerge when MPs in the required
amounts for a structural analysis by X-ray crystallography and nuclear
magnetic resonance (NMR) spectroscopy need to be prepared *(1,2)*. Unfortunately, only a very limited number of MPs could so far be produced in
conventional *Escherichia coli* in vivo expression systems at a level of at
least 1 mg/L of culture *(2,3)*.

In contrast, in vitro cell-free (CF) expression systems have recently been
shown to be considerably suitable for the high level expression of MPs *(4–
6)*. This finding might represent a breakthrough for the structural and functional analysis of MPs whenever high amounts of protein are required. The
advantage of CF expression systems in the production of MPs might be
owing to the elimination of some principal problems occurring in conventional in vivo systems, like toxicity of the overproduced MPs upon insertion into the cytoplasmic membranes, poor growth of overexpressing
strains, proteolytic degradation of the expressed MPs, or generally unfa-
vorable impacts on cellular metabolisms. In this chapter, we describe the
efficient production of MPs in a CF-coupled transcription/translation system using *E. coli* S30 cell extract and phage T7 RNA-polymerase for transcription. The set up of the reaction is in the continuous exchange mode,
where two compartments, holding the reaction mixture (RM) and the feeding mixture (FM), are separated by a semipermeable membrane *(7–9)*. Key
elements of the CF system like the bacterial S30-extract preparation *(10)*,
the energy system and the concentrations of precursors and of beneficial
additives have been optimized to yield several milligrams of recombinant
MP per one single milliliter of RM.

It should be highlighted that MPs can be produced either as a precipitate or as soluble proteins. The CF expression system is considerably
tolerant upon relatively high concentrations of a variety of detergents
and lipids *(4–6)*, and we compile suitable substances that can be provided as a hydrophobic environment to stabilize the synthesized MPs
immediately after translation. Alternatively, the vast majority of MPs
produced without supplemented detergents most likely will end up as
precipitates *(4,5)*. However, those precipitates usually can easily be solubilized with suitable detergents without the necessity to apply extensive
denaturation and renaturation steps as known from refolding protocols

of inclusion bodies. The organization of the CF produced MP precipitates might therefore be different from that of inclusion bodies formed during in vivo expressions.

## 2. Materials

### *2.1. S30-Extract Preparation*

1. 10-L Fermenter.
2. French Press cell disruption device.
3. Liquid $N_2$.
4. S30-A buffer: 10 m$M$ Tris-acetate, pH 8.2, 14 m$M$ Mg(OAc)$_2$, 6 m$M$ β-mercaptoethanol, 0.6 m$M$ KCl. Prepared as 50X stock solution and stored at 4°C.
5. S30-B buffer: 10 m$M$ Tris-acetate, pH 8.2, 14 m$M$ Mg(OAc)$_2$, 0.6 m$M$ KCl, 1 m$M$ DTT, 0.1 m$M$ phenylmethanesulfonyl fluoride. Prepared as 50X stock solution and stored at 4°C.
6. S30-C buffer: 10 m$M$ Tris-acetate, pH 8.2, 14 m$M$ Mg(OAc)$_2$, 0.5 m$M$ DTT, 0.6 m$M$ KOAc. Prepared as 50X stock solution and stored at 4°C.
7. NaCl, 4 $M$. Store at 4°C.
8. Terrific Broth (TB) medium (per liter): 24 g yeast extract, 12 g tryptone, 4 mL 100% glycerol, 100 m$M$ potassium phosphate buffer.
9. Dialysis tubes, type 27/32 MWCO 14 kDa (Roth, Karlsruhe, Germany; cat. no. 1784.1).
10. Bacterial strain A19 [*rna19 gdh A2 his95 relA1 spoT1 metB1*] *E. coli* Genetic Stock Center (*E. coli* Genetic Stock Center, New Haven, CT; CG, CGSC no. 5997).
11. Centriprep devices YM-10 (Amicon, Witten, Germany; cat. no. 4305).

### *2.2. Reaction for Cell-Free Protein Expression*

1. Total tRNA from *E. coli* (Roche Diagnostics GmbH, Mannheim, Germany; cat. no. 109550), 40 mg/mL in distilled water. Store at –20°C.
2. Pyruvate kinase, 10 mg/mL (Roche Diagnostics GmbH, cat. no. 109045). Store at –20°C.
3. T7-RNA-polymerase, 40 U/μL (Roche Diagnostics GmbH, cat. no. 881775) (*see* **Note 1**). Store at –20°C.
4. NTP-Mix, 75-fold, 90 m$M$ ATP, 60 m$M$ each CTP, GTP, UTP, pH 7.0 with NaOH. Store at –20°C.

5. 0.5 *M* Dithiothreitol. Store at −20°C.

6. 20 m*M* Folinic acid. Store at −20°C.

7. PEG 8000, 40%. Store at −20°C.

8. $NaN_3$, 10%. Store at 4°C.

9. Amino acid mixture, 4 m*M* of each of the 20 amino acids (*see* **Note 2**). Store at −20°C.

10. Amino acid mixture R, C, W, M, D, E, 16.7 m*M* (*see* **Note 3**). Store at −20°C.

11. RNAguard porcine RNase inhibitor, 40 U/μL (Amersham, cat. no. AP 27-0816-01). Store at −20°C.

12. Complete mini protease inhibitor (Roche Diagnostics GmbH, cat. no. 1836153), 50-fold concentrated in water. Store at −20°C.

13. Acetyl phosphate, potassium salt, 1 *M* in distilled water, pH 7.0 with KOH. Store at −20°C.

14. Phosphoenol-pyruvic-acid, mono-potassium salt, 1 *M* in distilled water, pH 7.0 with KOH. Store at −20°C.

15. HEPES buffer, 2.5 *M*, pH 8.0 with KOH. Store at 4°C.

16. 4 *M* KOAc. Store at 4°C.

17. 1 *M* $Mg(OAc)_2$. Store at 4°C.

18. 10 *M* KOH.

19. 5 *M* NaOH.

20. Reaction container for the cell-free expression: microdialysers (cutoff 15 kDa, Roth, cat. no. T698.1), and dispodialysers (reaction volume 1 mL or higher, Roth, cat. no. O359.1). Devices may be reused several times (*see* **Note 4**).

21. Incubator for the cell-free reaction, e.g., standard shaker with temperature control or rollers like the Universal Turning Device (Vivascience, Göttingen, Germany; cat. no. IV-76001061) placed in an incubator (*see* **Note 5**).

22. Plasmid vectors containing the T7 promoter regulatory region, e.g., pET21 (Merck Biosciences, Darmstadt, Germany).

23. Glass vials, 50 mL (Roth, cat. no. X663.1).

## 2.3. Preparation of Solubilized or Reconstituted Membrane Proteins

1. Detergents: α-[4-(1,1,3,3-tetramethylbutyl)-phenyl]- ω-hydroxy-poly(oxy-1,2-ethandiyl) (Triton X-100), digitonin, dodecylpoly(ethyleneglycolether)$_9$ (Thesit), sodium dodecyl sulfate, polyethylenglycododecylether (Brij-35), n-octyl-β-D-glucopyranosid (β-OG); (Sigma-Aldrich, Taufkirchen, Ger-

many). 3-[*N*-(cholanamidopropyl)-dimethyl-ammonio]-1-propanesulfonate (CHAPS), Polyoxy-ethylene (20) sorbitane monolaurate (Tween-20; Roth). n-dodecylphosphocholine (DPC), 1-myristoyl-2-hydroxy-sn-glycero-3-[phosphor-rac-(1-glycerol)] (LMPG); (Avanti-Lipids, Alabaster, AL).
n-dodecyl-β-ᴅ-maltoside (DDM); (Glycon Biochemicals GmbH Luckenwalde, Germany).

2. *E. coli* lipid mixture (Avanti-Lipids) stock suspension, 50 mg/mL in distilled water.
3. BT Bio-beads SM-2 (Bio-Rad, München, Germany).

## 3. Methods

### 3.1. Construction of Expression Plasmids

The transcription of the target gene depends on the T7 RNA-polymerase and therefore the essential T7 regulatory sequences (promoter, ε-enhancer or g10 sequence, ribosomal binding site, terminator) need to flank the coding region (*see* **Fig. 1**). Standard expression vectors like that of the pET series can be used for cell-free expression (*see* **Note 6**).

### 3.2. Preparation of the S30 Cell-Free Extract

The preparation of the S30-extract includes fermentation and harvesting of the cells, the cell disruption, run-off procedure and dialysis. Although we recommend to finish the complete process consecutively, the protocol could optionally be interrupted after the harvesting step (*see* **Note 7**). The quality of individual batches of S30-extracts can vary considerably and at least an optimization for the $Mg^{2+}$ and $K^+$ ion concentrations should be made for each new preparation.

1. Take a fresh overnight culture from the *E. coli* strain A19 in TB medium and inoculate 10 L of TB medium in a fermenter at a ratio of 1:100 (*see* **Note 8**).
2. Incubate the cells at 37°C with vigorous stirring and good aeration.
3. Control the bacterial growths by measuring the optical density at 595 nm until the cells reach the mid-log phase corresponding to an $OD_{595}$ of approx 3.5 (*see* **Note 9**).
4. Switch off the heating of the fermenter and chill down the broth to ≤10°C as quickly as possible (≤45 min) to stall the growth of the bacteria (*see* **Note 10**). The final $OD_{595}$ of the broth after cooling should be ≤4.5.

promoter                              ε-enhancer                    RBS

TAATACGACTCACTATA - N$_{38}$ - TTAACTTTA – N$_2$ - AAGGAG – N$_7$ -

coding sequence                                        terminator

ATG – N$_x$ – TAA –  N$_x$ – AACCCCTTGGGGCCTCTAAACGGGTCTTGAGGGGTTTTT

Fig. 1. Essential T7 regulatory sequences for optimal cell-free protein production.

5. Harvest the cells by centrifugation at 4°C for 10 min at 7000*g* in precooled beakers.
6. Resuspend the pellet with a glass-rod or similar in 100 mL of S30-A buffer precooled to 4°C. Centrifuge at 8000*g* at 4°C for 10 min. Repeat the washing step two more times while the final centrifugation step should last 30 min. Out of a 10-L fermenter with TB medium, you should now yield approx 60–70 g wet-weight of bacterial cells.
7. Suspend the cell pellet in an equal volume (e.g., for 65 g cells use 65 mL of buffer) of S30-B buffer precooled at 4°C.
8. Disrupt the cells by passing through a French-Press cell precooled at 4°C with one pass at 1200 psi (*see* **Note 11**).
9. Pellet cell debris by centrifugation at 30,000*g* at 4°C for 30 min. The pellet can be quite smooth at this stage and a turbid solution immediately above the pellet, if present, should        not be transferred.
10. Transfer the upper two-thirds of the supernatant in a fresh vial and repeat the centrifugation step.
11. Remove the upper two-thirds of the supernatant carefully, adjust to a final concentration of 400 m*M* NaCl and incubate at 42°C for 45 min in a water bath (*see* **Note 12**). The solution will become turbid.
12. Fill the turbid solution in a dialysis tubing (cut-off 14 kDa) and dialyse at 4°C against 60 vol of S30-C buffer with gentle stirring. Exchange the dialysis buffer after 2 h and continue to dialyse overnight.
13. Centrifuge the extract at 30,000*g* at 4°C for 30 min.
14. Remove the clear supernatant, fill suitable aliquots (*see* **Note 13**) in plastic tubes and freeze in liquid nitrogen. The frozen extract can be stored at –80°C for months. One 10-L fermenter yields approx 50 ml 60 mL of extract.

15. Optional step: Before aliquoting, the extract could be concentrated to approximately half of the original volume by using microconcentrator devices with a cut-off of 10 kDa (*see* **Note 14**).

## 3.3. Setup of a Cell-Free Reaction

Generally, the reaction can be set up in analytical modes for optimization and screening reactions as well as in preparative modes for the production of milligram amounts of recombinant protein. As a device for analytical scale reactions with a RM volume of 70 µL, microdialysers that are commercially available with different MWCOs can be used (*see* **Fig. 2A** and **Note 15**). For preparative reactions with RM volumes of 500 µL and larger dispodialysers (or simple dialysis tubes) can be used (*see* **Fig. 2B**). The ratio of RM/FM should be 1:14 (microdialysers) or 1:17 (dispodialysers) (v/v) (*see* **Note 16**). The final concentrations of the individual components are listed in **Table 1**. An example of a pipetting protocol is given for a preparative reaction with 1 mL RM and 17 mL of FM in **Table 2**.

1. Thaw aliquoted stock solutions and mix carefully. The enzymes, tRNA and the S30 extract should be kept on ice after thawing.
2. First make the master-mix FRM by pipetting the components common to FM and RM (*see* **Table 2**).
3. Take the appropriate aliquots from the master-mix FRM (*see* **Table 2**) and first complete the FM.
4. Preincubate the FM at 30°C in a water bath.
5. Start completing the RM (*see* **Table 2**). All steps should be carried out on ice. Mix but do not vortex. Keep the solution on ice.
6. Fill 17 mL of the FM into the FM-compartment. (*see* **Note 17**).
7. Fill 1 mL of RM into the RM-compartment. Avoid any air-bubbles at the dialysis membrane that might restrict an efficient exchange between the two compartments. All dialysers should be thoroughly rinsed with distilled water before use.
8. Incubate the reaction on a suitable shaking or rolling device, e.g., the Universal Turning Device with at least 40 rpm at 30°C for 15 h to 20 h.
9. Harvest the RM containing the recombinant protein. Depending on the cut-off of the dialysis membrane, the recombinant protein might also be present in the FM.

**A**

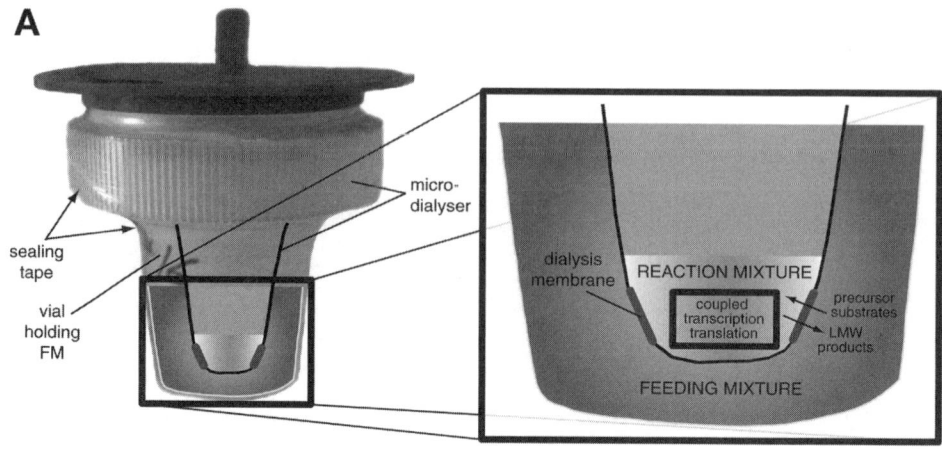

**B**

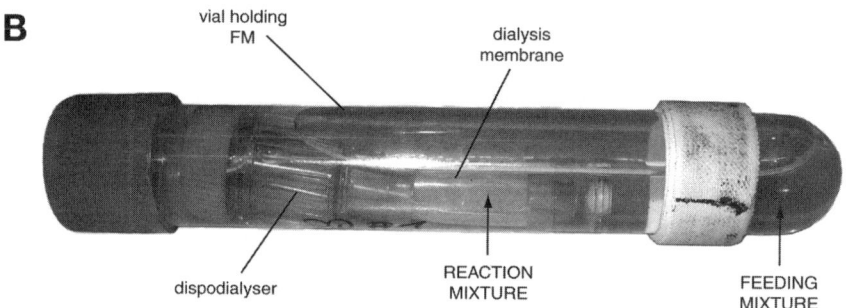

Fig. 2. Design of cell-free reaction containers. (**A**) Microdialyser for analytical scale reactions. The microdialyser holding the approx 70 µL of reaction mixture (RM) is placed into a suitable vial (e.g., bottom part of a 50-mL Falcon plastic tube) holding the 1 mL of feeding mixture (FM). The vial is fixed to the microdialyser by a sealing tape (e.g., nescofilm). The setup is than incubated on a shaker at 30°C. (**B**) Dispodialyser for preparative scale reactions. A dispodialyser holding a suitable volume of RM is placed into a plastic vial containing the corresponding volume of FM. The setup is then incubated on a roller at 30°C.

## 3.4. Optimization of the Expression Yields

This step is especially important to achieve the highest yields possible, i.e., incorporation efficiencies of the added amino acids of up to 20%. We recommend to run an optimization experiment at least for the highly critical concentrations of $Mg^{2+}$ and $K^+$ ions with each newly pre-

**Table 1**
**Protocol for Cell-Free Protein Expression**

| Component | Stock concentration | Final concentration in RM |
|---|---|---|
| S30-extract/S30-C-buffer | 100% | 35% |
| Plasmid DNA | 0.3 mg/mL | ≥15 μg/mL |
| RNAguard | 39.8 U/μL | 0.3 U/μL |
| T7 RNA-polymerase | 40 U/μL | ≥3 U/μL |
| Escherichia coli tRNA | 40 mg/mL | 500 μg/mL |
| Pyruvate kinase | 10 mg/mL | 40 μg/mL |

| Component | Stock concentration | Final concentration in RM + FM |
|---|---|---|
| Amino acids | 4 mM/16.7 mM | 1 mM |
| Acetyl phosphate | 1 M | 20 mM |
| Phosphoenol-pyruvic-acid | 1 M | 20 mM |
| ATP | 360 mM | 1.2 mM |
| CTP, GTP, UTP | 240 mM | 0.8 mM each |
| 1.4-Dithiothreitol (DTT) | 500 mM | 2 mM |
| Folinic acid | 20 mM | 0.2 mM |
| Complete protease inhibitor | 50X (1 tablet/0.2 mL) | 1 tablet /10 mL |
| HEPES-KOH pH 8.0 | 2.5 M | 100 mM |
| Magnesium acetate | 1 M | 14 mM |
| Potassium acetate | 4 M | 290 mM |
| Polyethylenglycol 8000 | 40% | 2% |
| Sodium azide | 10% | 0.05% |

Concentrations of $Mg^{2+}$ and $K^+$ are highly critical and should be subject of optimization. 9.1 mM magnesium acetate and 150.8 mM potassium acetate are added, 4.9 mM $Mg^{2+}$ results from the S30-extract and 139.2 mM $K^+$ results from other reaction components. Amino acids and T7-RNA-polymerase are limiting compounds and only minimal concentrations are given. The standard amino acid concentrations according to the protocol will be 0.5 mM in the RM and 1 mM in the FM.

**Table 2**
**Pipetting Protocol for a 1-mL Cell-Free Reaction**

| Stock solution | Master-mix FRM |
|---|---|
| 10% NaN$_3$ | 92 µL |
| 40% PEG8000 | 918 µL |
| 4 $M$ KOAc | 692 µL |
| 1 $M$ Mg(OAc)$_2$ | 167 µL |
| 25X Buffer | 646 µL |
| 50X Complete + EDTA | 367 µL |
| 20 m$M$ Folinic acid | 184 µL |
| 0.5 $M$ DTT | 73 µL |
| 75X NTP-mix | 245 µL |
| 1 $M$ PEP | 367 µL |
| 1 $M$ AcP | 367 µL |
| 4 m$M$ AA-mix | 2295 µL |
| 16.7 m$M$ RCWMDE-mix | 1099 µL |
| | 7513 µL |

| Stock solution | FM | RM |
|---|---|---|
| Master-mix FRM | 6957 µL | 409 µL |
| 4 m$M$ AA-Mix | 2125 µL | – |
| S30-C-buffer | 5950 µL | – |
| 10 mg/mL Pyruvate kinase | – | 4 µL |
| 40 mg/mL *Escherichia coli* tRNA | – | 13 µL |
| 40 U/µL T7-RNA-Polymerase. | – | 75 µL |
| 39.8 U/µL RNAsin | – | 8 µL |
| S30-Extract | – | 350 µL |
| 0.3 mg/mL Plasmid-DNA | – | 50 µL |
| H$_2$O | 1968 µL | 91 µL |
| | 17,000 µL | 1000 µL |

All volumes less than 1 µL have been rounded up.

pared batch of S30-extract. The final Mg$^{2+}$ and K$^+$ concentrations should be varied by addition of suitable amounts of Mg(OAc)$_2$ or KOAc in the range between 12 and 17 m$M$ and 270 and 330 m$M$, respectively (*see* **Note 18**). The Mg$^{2+}$ concentrations should be titrated in 1 m$M$ steps and the K$^+$ concentrations in 10-m$M$ steps. Consider that the optima of the two ions depend on each other.

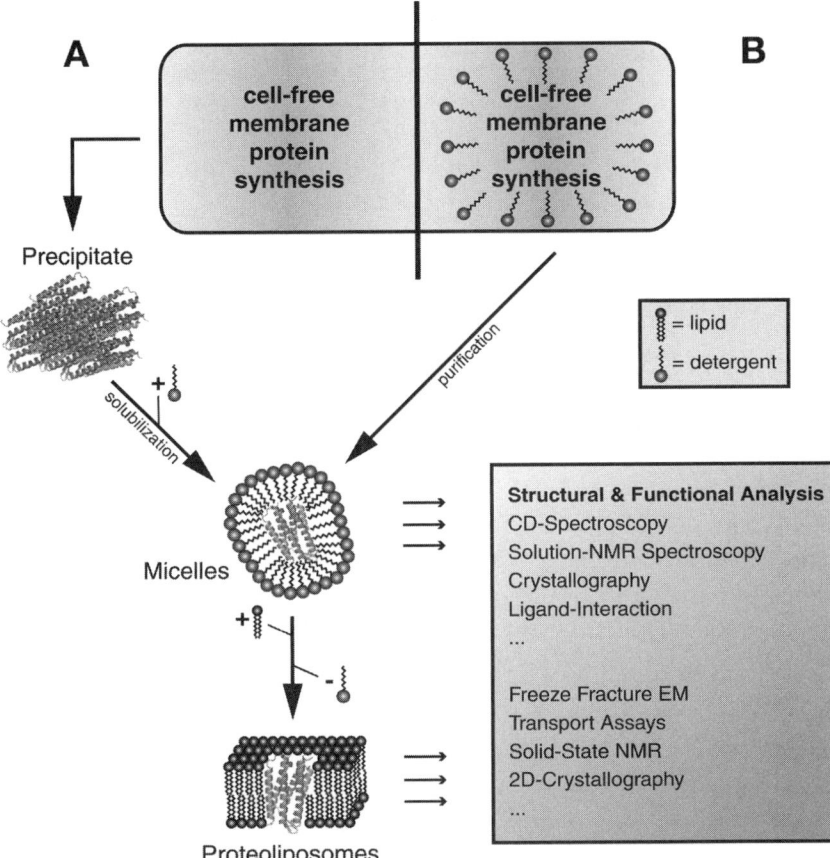

Fig. 3. Different modes of membrane protein (MP) production by cell-free (CF) expression. (**A**) Production of MP precipitates followed by resolubilization after addition of detergents. (**B**) Production of soluble MPs by addition of detergents directly into the CF system. The MPs solubilized in micelles by the two different modes can be directly analyzed or reconstituted into liposomes for further analysis.

## 3.5. Cell-Free Expression of Membrane Proteins as a Precipitate

MPs can be expressed in the cell-free system either as a precipitate or as soluble proteins by the addition of suitable detergents or lipids directly into the reaction (*see* **Fig. 3**). Although it might be more likely to obtain higher yields if a MP is produced as a precipitate, preparative amounts of

| Origin: | procaryotic | procaryotic | procaryotic | procaryotic | eucaryotic |
|---|---|---|---|---|---|
| **Type**: | α-helical | α-helical | α-helical | ß-barrel | α-helical |
| **Function**: | Multi-Drug Transporter | Amino acid Transporter | Heavy-Metall Resistance | Nucleoside Transporter | Vasopressin Receptor |
| **Family**: | SMR | RhtB | TDT | OMP | GPCR |
| **MP**: | SugE   EmrE | YfiK | TehA | Tsx | V2-R |

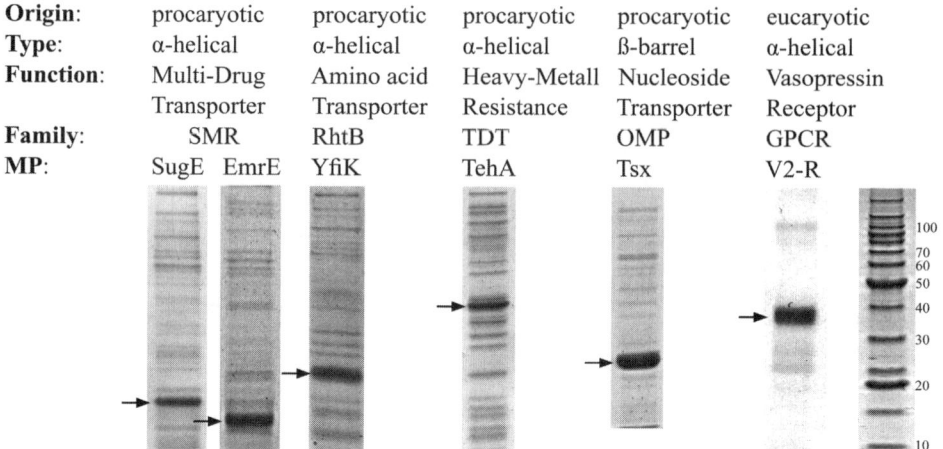

Fig. 4. Membrane proteins (MPs) from different families that have been produced as a precipitate by cell-free expression. The precipitate of an analytical reaction (70 μL of reaction mixture) was suspended in 70 μL of buffer and the proteins in 1 μL of each suspension were separated by electrophoresis in a 12% sodium dodecyl sulfate-polyacrylamide-gel. The arrows indicate the overproduced recombinant MPs. The right lane shows the molecular weight marker. GPCR, G-protein coupled receptor; OMP, outer membrane protein; SMR, small multidrug transporter.

protein can generally be synthesized by both ways. The precipitates are predominantly formed by the target protein but they still can contain a considerable variety of coprecipitated proteins originating from the S30-extract (*see* **Fig. 4**). The cell-free production of MPs in form of a precipitate follows exactly the standard protocol (*see* **Subheading 3.3.**). The synthesis of the MP can be followed by the increasing turbidity of the RM. After incubation, the MP can be harvested by centrifugation of the RM at 10,000*g* for 10 min.

## 3.6. Solubilization of CF-Expressed MPs

Interestingly, the precipitated MPs do not seem to be completely unfolded if compared with the inclusion body formation by conventional in vivo expression. The solubilization of the cell-free produced MP precipitates usually does not require extensive unfolding and refolding steps like the treatment with guanidinium hydrochloride or urea. Most MP precipitates start to solubilize more or less immediately after addition of a suitable detergent. The choice of the detergent certainly depends on the nature of the recombinant MP. A list of detergents that have been useful for solubilization is given in **Table 3**. It is important to mention that some detergents like, e.g., LMPG, that are only fairly suitable for the soluble expression of MPs, are on the other hand highly efficient in the solubilization of MP-precipitates. However, a 100% efficiency of solubilization might never been achieved.

1. Centrifuge the RM containing the suspended MP-precipitate at 20,000$g$ for 10 min at room temperature.
2. Discard supernatant and wash the precipitate in an adequate solubilization buffer (e.g., 50 m$M$ Na$_2$HPO$_4$, pH 7.8, 1 m$M$ DTT) of a volume equal to the RM (*see* **Note 19**). The pellet should be carefully suspended by pipetting.
3. Repeat centrifugation step.
4. Suspend the pellet in solubilization buffer in a volume equal to the RM volume and containing detergent (e.g., 2% DPC [w/v]). Incubate at 30 – 37°C up to several hours (*see* **Note 20**).
5. Remove unsoluble protein by centrifugation at 20,000$g$ for 10 min. at room temperature and remove the solubilized MP in the supernatant for further analysis.

### *3.7. Reconstitution of Solubilized MPs*

The successful reconstitution strongly depends on the right choice of a suitable lipid or lipid mixture and those should be primarily selected according to the origin of the recombinant MP. *E. coli* lipids are predominantly composed of phosphatidylethanolamine (PE), phosphatidylglycerol, and cardiolipin, whereas eukaryotic lipids are mainly mixtures of phosphatidylcholine, cholesterol, PE, and phosphatidylserine. A huge variety of defined lipids or lipid mixtures as well as crude lipid isolates of different origin are commercially available for selection. As a rapid control of an effective and homogeneous reconstitution of MPs we recommend the analysis of the proteoliposomes by freeze-fracture electron microscopy. The homogeneous insertion of protein particles into the membrane as shown in **Fig. 5** for the cysteine transporter YfiK provides first evidence of a structural and functional reconstitution of the MP. As an

**Table 3**
**Detergents Used for the Solubilization of Cell-Free Expressed MPs**

| Detergent | Nature | Mass (Da) | CMC (mM) | Concentration (x CMC) | Solubilization of precipitates | Soluble MPs Em | Tsx | V2R |
|---|---|---|---|---|---|---|---|---|
| SDS | A | 288.38 | 2.6 | 2 | +++ | 0 | 0 | 0 |
| Tween-20 | N | 1228 | 0.059 | 10 | + | 0 | 0 | 0 |
| CHAPS | Z | 614.9 | 8.0 | 1.5 | ± | 0 | I | 0 |
| β-OG | N | 292.4 | 19 | 2 | ± | I | I | 0 |
| DPC | Z | 351.5 | 1.5 | 1 | ++ | I | 0 | I |
| Thesit | N | 583 | 0.1 | 17 | + | I | I | I |
| LMPG | A | 478.5 | 0.05 | 4 | +++ | I | I | I |
| Triton X-100 | N | 647 | 0.23 | 15 | + | I | II | I |
| DDM | N | 348.5 | 0.19 | 10 | + | II | II | I |
| Digitonin | N | 1229.31 | 0.73 | 2 | + | II | II | II |
| Brij-35 | N | 1199.57 | 0.08 | 10 | + | III | III | III |

The highest concentrations (x CMC) that have been added into the CF reaction are shown. For the solubilization of precipitates, considerably higher concentrations can be used. A, anionic; N, nonionic; Z, zwitterionic; CMC, critical micellar concentration. The production of soluble MP is compared with the yields obtained from the production as precipitates in standard CF reactions without detergent. 0, no soluble MP production; I, low level soluble MP production ≤20%; II, medium level soluble MP production between 20 and 50%; III, high level soluble MP production ≥50%. Em, *Escherichia coli* α-helical multidrug transporter EmrE; Tsx, *E. coli* β-barrel transporter Tsx; V2R, porcine vasopressin receptor.

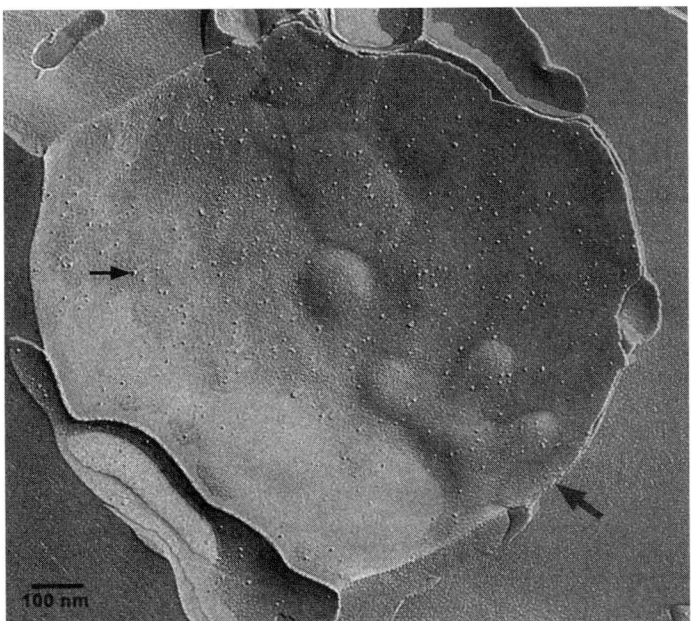

Fig. 5. Freeze-fracture electron microscopy of the cell-free expressed *Escherichia coli* cysteine transporter YfiK after reconstitution into an *E. coli* lipid mixture. The YfiK protein was first produced as a precipitate in a standard cell-free reaction, then solubilized in 1% (v/v) LMPG and subsequently reconstituted into *E. coli* lipid vesicles by the above described procedure. The randomly distributed particles (small arrow) indicate the homogenous in vitro incorporation of the membrane protein into the vesicular membranes (bold arrow). (Courtesy of W. Haase.)

example, we describe the reconstitution of a CF produced MP in an *E. coli* lipid mixture.

1. Add *E. coli* lipids (50 mg/mL in water) in a molar ratio of 2000:1 (lipid:MP) to the solubilized protein.
2. Incubate at 30°C for 1 h.
3. Wash 750 mg biobeads in 10 mL of 100% methanol, let the biobeads settle down and discard the supernatant.
4. Wash the biobeads three times with 10 mL of distilled water.
5. Finally wash the biobeads two times in the buffer used for the detergent solution.

6. Presaturation with lipids: add 100 µL of the lipid suspension to 1 mL of washed biobeads in 10 mL of detergent buffer (e.g., 25 m*M* HEPES, pH 7.4, 150 m*M* NaCl) and incubate for 30 min at RT.
7. Add lipid-presaturated biobeads in a ratio biobeads:detergent of 100:1 (w/ w) to a vial containing the micelle suspension and let the detergent adsorb to the biobeads by incubation on a shaker at 30°C.
8. Incubate overnight on a shaker at 30°C.
9. Next morning, remove the vial from the shaker, let the biobeads settle down and transfer the supernatant into a new vial. Then incubate with a second aliquot of lipid-presaturated biobeads for 6 h at 30°C on a shaker.
10. Harvest the supernatant containing the proteoliposomes. The proteoliposomes can be stored at 4°C until or frozen in liquid nitrogen until further analysis.

## 3.8. Cell-Free Expression of Soluble Membrane Proteins

1. The expression of soluble MPs in the preparative scale can require first some optimization steps to evaluate the optimal reaction conditions. At the beginning, only some micrograms of protein might be produced. We therefore recommend the addition of an antigen epitope tag like the T7-tag at the N-terminal end of the recombinant protein to facilitate the detection of the expressed MP in Western blots. Furthermore, an additional poly(His)$_6$-tag at the C-terminal end of the recombinant protein would enable the fast purification out of the RM by standard metal-chelate-chromatography .
2. The cell-free expression system is considerably tolerant against a variety of detergents even if supplied in relatively high concentrations *(4–6)*. Detergents that are compatible with the CF synthesis can be either ionic or nonionic, but in most cases they have a relatively low critical micellar concentration. First, the most appropriate detergents have to be identified as their effects on the expression and solubility of specific MPs can be very different (*see* **Fig. 6**). Overall, Brij-35 has so far turned out to be one of the most effective detergents for the production of soluble MPs (*see* **Table 3** and **Fig. 6**). A detergent screen including a variety of detergents (*see* **Table 3**) should be set up in the analytical scale mode. An example for the CF expression of the nucleoside transporter Tsx in presence of different detergents is shown in **Fig. 6**. In that case, Brij-35, Triton X-100, and DDM would clearly be the best choices to produce soluble protein. Also the addition of mixed micelles composed of two or more detergents might be considered. The production of MPs can be monitored by

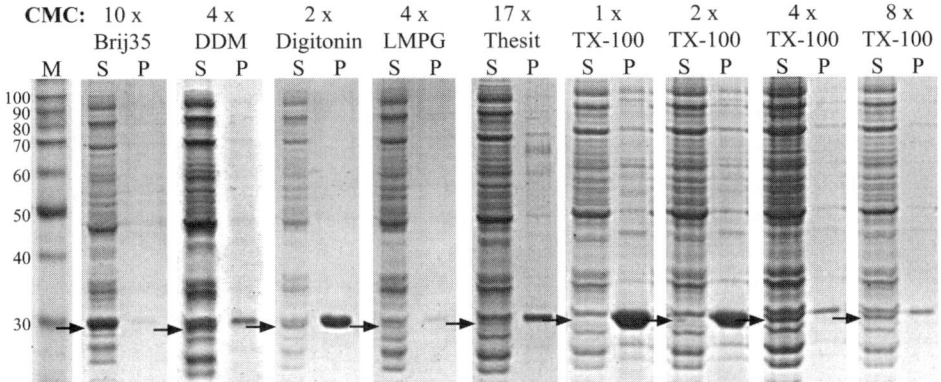

Fig. 6. Soluble cell-free (CF) expression of the β-barrel-like nucleoside transporter Tsx in presence of detergents. Samples were from analytical scale CF reactions with 70 μL reaction mixture (RM). The concentrations of the added detergents are given in critical micellar concentration. After the reaction, 0.8 μL of the RM, or of the precipitated proteins suspended in a volume equal to the RM volume, were loaded on a 12% sodium dodecyl sulfate-polyacrylamide-gel. The arrows indicate the soluble Tsx protein in the reaction mixture. S, soluble protein; P, precipitated protein; M, marker proteins (kDa); TX-100, Triton X-100.

Coomassie-Blue staining or Western blotting after separation of the RM by sodium dodecyl sulfate-polyacrylamide gel-electrophoresis. To differentiate between soluble MPs and precipitate, the RM should be fractionated by centrifugation at 20,000g for 30 min at RT prior to analysis.

3. After the most suitable detergents for your specific protein have been identified, run a second series of analytical scale reactions to define the optimal concentration of the preferred detergent. For example, the highest yields of soluble Tsx protein in presence of Triton X-100 were only obtained at concentrations of 4X critical micellar concentration and above (*see* **Fig. 6**).

4. The solubilized MPs in micelles might not necessarily adopt a structured and functional conformation. A rapid first analysis of the MP structure should therefore be carried out to verify the structured folding of the CF expressed MP. Suitable techniques might be Circular Dichroism spectroscopy or the recording of heteronuclear single quantum correlation spectra by solution NMR spectroscopy. Highly valuable would be certainly the availability of any activity or binding assay.

## 4. Notes

1. T7 RNA-polymerase will be the most expensive component of the reaction and in addition, commercial enzymes are often too low concentrated. At least 3 U final concentration of T7 RNA-polymerase per microliter of RM should be added but up to 40 U/µL can be used. We therefore highly recommend the overproduction of the enzyme in *E. coli (11)*. We isolate the T7-RNA-polymerase from strain BL21 (DE3) by a single-step purification with a Q-sepharose column. Out of a 4-L fermentation, the yield can be approx $5 \times 10^5$ U. The isolated T7 RNA-polymerase can be stored in glycerol at –80°C for many months.

2. L-tyrosine is prepared as 20 m$M$, the other 19 amino acids are made as 100 m$M$ stock solutions dissolved in water. The solubility of L-aspartic acid, L-cysteine, L-glutamic acid, and L-methionine could be improved by using 100 m$M$ HEPES pH 7.4. L-tryptophan is dissolved in 100 m$M$ HEPES, pH 8.0, upon sonication in a water bath. Appropriate aliquots of the individual stocks are then combined in the amino acid mixture. The final concentration of amino acids in the reaction is 1 m$M$. The yields of recombinant protein might be improved by using higher concentrations, and by adjusting the amino acid concentrations according to the composition of the recombinant protein. Least abundant amino acids (present ≤3% in the protein) should then be added at 1.25 m$M$, medium abundant (between 3 and ≤8%) at 1.8 m$M$ and highly abundant (more than 8%) at 2.5 m$M$ final concentration.

3. Some amino acids tend to be unstable and increasing their concentration in the reaction significantly improves the synthesis of recombinant proteins *(12,13)*.

4. Dialysers containing membranes made of regenerated cellulose should be preferred as they enable a better exchange of compounds if compared with cellulose-ester. Dialysers might be reused after washing with distilled water and they can be stored in distilled water supplemented with 0.01% sodium azide at 4°C.

5. Depending on the design of the reaction container, quite a variety of incubators can be used. The only consideration is that temperature control and an efficient agitation of the reaction device, either by rolling, shaking, or stirring, can be provided.

6. RNase in the lysis buffer during plasmid isolation principally could cause problems in the cell-free reaction and it might be avoided during DNA preparation. Distilled water should be used to finally dissolve the DNA to prevent the addition of undesired ions into the cell-free reaction. The plas-

mid DNA should be concentrated in a speedvac to a final concentration of at least 0.1 µg/µL. The final concentration of DNA in the cell-free reaction should be at least 15 µg/1 mL of RM.

7. At this stage, the cells could be frozen at –80°C as a thin plate wrapped in a sheet of aluminum foil and stored until further usage. However, in our hands the quality of the S30 extract was always considerably improved if the preparation was consecutively completed without interruption.

8. A variety of bacterial strains including *E. coli* A19, D10, or BL21 has been described as source for cell-free S30-extracts. In our hands, A19 yielded the most efficient extracts with the best reproducible quality. Using a fermenter to grow the bacteria might be essential because of the better supply of oxygen.

9. Harvesting the cells in the mid-log phase is essential. For new modifications of the fermentation conditions, an initial pilot experiment should always be carried out to define the exact optical density of the culture at mid-log phase.

10. It is critical to cool down the broth as quickly as possible. As the cells rapidly divide, too slow cooling might result in entering the late-log or even stationary phase of growth and that could considerably reduce the quality of the extract. For the described conditions, the reduction of temperature from 37 to 10°C should be optimally completed in not more than 45 min. An external cooling-unit could be connected or the broth in the fermenter could be cooled down by adding blocks of frozen TB medium.

11. Using a French Press device for cell disruption is important. Disruption by ultra-sonification would result in bad quality of the extract, probably by disintegration of the ribosomes.

12. This step causes the dissoziation of endogenous mRNA from the ribosomes. Other run-off procedures by adding substrates to terminate transcriptions have been described (*10,14*) and are similar effective, but more expensive. The incubation at the relatively high heat shock temperature of 42°C has been proven to be essential. Using lower temperatures like 37°C are by far not as effective. A lot of proteins in the extract precipitate during this step while the proteins necessary for the translation process remain stable.

13. Make aliquots of different appropriate sizes according to your requirements (e.g., 100 µL, 500 µL, 1 mL) as the individual aliquots must not be refrozen.

14. Condensing the extract twofold can increase the efficiency to approx 1.5 times. The final protein concentration in the extract should be between 20 and 30 mg/mL.

15. Relatively large cut-offs up to 50 kDa can be used without having significant leaching of the proteins relevant for the transcription/translation process. It is therefore assumed that those proteins form a large macromolecular complex together with the mRNA. The only protein that is released from that complex is the synthesized recombinant target protein. The selection of the MWCO of the reaction device should therefore depend on the MW of the target protein and whether it should remain in the RM. Generally, MWCOs <10 kDa tend to reduce the expression yields probably owing to a more restricted exchange of low molecular weight substances and we recommend to routinely use a MWCO of 15 kDa.

16. The ratio RM/FM is certainly important for an efficient production of recombinant protein. The higher the volume of the FM, the more protein can be produced. However, the FM contains relatively expensive precursors like PEP, acetyl phosphate, nucleotides, or even labeled amino acids. We found the given ratio to be an optimal compromise to yield high levels of protein at reasonable costs. The exchange of the FM after several hours of incubation would additionally increase the yield of the recombinant protein. However, as the increase usually is only approx 30%, it would be more economically to run two identical reactions separately.

17. For preparative scale reactions, e.g., in dispodialysers, standard glass vials of suitable sizes can be used as FM compartment. For analytical scale reactions, e.g., in microdialysers, suitable plastic containers like the lower part of 50-mL polypropylene test tubes (Greiner, Solingen, Germany, cat. no. 210261) that can be fixed to the microdialysers with a tape (e.g., Nescofilm, Azwell, Osaka, Japan) can be used.

18. For the calculation of the final concentrations, it has to be considered that 5 m$M$ of Mg$^{2+}$ and 140 m$M$ of K$^+$ are already present in the reaction from other components.

19. It could be beneficial to include detergents that are not suitable for dissolving the recombinant protein into the washing buffer. Coprecipitated contaminants might then be dissolved, resulting in a purer sample of the MP.

20. Considerably higher detergent concentrations can be used for the resolubilization of precipitated MPs if compared with the direct addition of detergent into the CF reaction for the production of soluble MPs.

## Acknowledgments

We are grateful to Vladimir Shirokov, Alexander Spirin, and Heinz Rüterjans for their help in setting up the cell-free expression system. We further thank Winfried Haase for the electron microscopical analysis of proteoliposomes.

## References

1. Arora, A. and Tamm, L. K. (2001) Biophysical approaches to membrane protein structure determination. *Curr. Opin. Struct. Biol.* **11**, 540–547.
2. Wang, D. N., Safferling, M., Lemieux, M. J., Griffith, H., Chen, Y., and Li, X. D. (2003) Practical aspects of overexpressing bacterial secondary membrane transporters for structural studies. *Biochim. Biophys. Acta* **1610**, 23–365.
3. Tate, C. G. (2001) Overexpression of mammalian integral membrane proteins for structural studies. *FEBS Lett.* **504**, 94–98.
4. Klammt, C., Löhr, F., Schäfer, B., et al. (2004) High level cell-free expression and specific labeling of integral membrane proteins. *Eur. J. Biochem.* **271**, 568–580.
5. Elbaz, Y., Steiner-Mordoch, S., Danieli, T., and Schuldiner, S. (2004). In vitro synthesis of fully functional EmrE, a multidrug transporter, and study of its oligomeric state. *Proc. Natl. Acad. Sci.* USA **101**, 1519–1524.
6. Berrier, C., Park, K. H., Abes, S., Bibonne, A., Betton, J. M., and Ghazi, A. (2004). Cell-free synthesis of a functional ion channel in the absence of a membrane and in the presence of detergent. *Biochemistry* 43, 12,585–12,591.
7. Baranov, V. I. and Spirin, A. S. (1993) Gene expression in cell-free system on preparative scale. *Methods Enzymol.* **217**, 123–142.
8. Spirin, A. S., Baranov, V. I., Ryabova, L. A., Ovodov, S. Y., and Alakov,Y. B. (1988) A continuous cell-free translation system capable of producing polypeptides in high yield. *Science* **242**, 1162–1164.
9. Kim, D. M. and Choi, C. Y. (1996) A semicontinuous prokaryotic coupled transcription/translation system using a dialysis membrane. *Biotechnol. Prog.* **12**, 645–649.
10. Zubay, G. (1973) *In vitro* synthesis of protein in microbial systems. *Annu. Rev. Genet.* **7**, 267–287.
11. Li, Y., Wang, E., and Wang, Y. (1999) A modified procedure for fast purification of T7 RNA-polymerase. *Protein Expr. Purif.* **16**, 355–358.
12. Kim, D. M. and Swartz, J. R .(1999) Prolonging cell-free protein synthesis with a novel ATP regeneration system. *Biotech. Bioengineering* **66**, 180–188.

13. Kim, D. M. and Swartz, J. R. (2000) Prolonging cell-free protein synthesis by selective reagent additions. *Biotech. Prog.* **16,** 385–390.

14. Kigawa, T., Yabuki, T., Matsuda, N., et al. (2004) Preparation of *Escherichia coli* cell extract for highly productive cell-free protein expression. *J. Struct. Funct. Genomic* **5,** 63–68.

# 4

# SIMPLEX: Single-Molecule PCR-Linked In Vitro Expression

*A Novel Method for High-Throughput Construction and Screening of Protein Libraries*

## Suang Rungpragayphan, Tsuneo Yamane, and Hideo Nakano

### Summary

A novel strategy for construction of protein libraries called "SIMPLEX: single-molecule PCR-linked in vitro expression" is described. A pool of genes is prepared and thereafter extensively diluted to give one molecule of DNA per well. Each individual molecule is separately amplified by PCR (single-molecule PCR) yielding a one-well-one-gene PCR library. Subsequently, the PCR library is directly transformed into a protein library by means of in vitro-coupled transcription/translation in an array format. Individual proteins in the library can be screened for target functions directly without further purification. The generated protein library is compatible with various selection methods. The strategy provides high-throughput construction and screening of protein libraries, and suits automation.

**Key Words:** SIMPLEX; single-molecule PCR; in vitro expression; coupled transcription/translation; PCR library; protein library; protein engineering; high-throughput screening.

From: *Methods in Molecular Biology*, vol. 375,
*In Vitro Transcription and Translation Protocols, Second Edition*
Edited by: G. Grandi © Humana Press Inc., Totowa, NJ

## 1. Introduction

Proteins with certain properties such as high activity, thermal stability, high selectivity, high affinity, and so on, some of which are not available from natural source, are often required in industrial or medical use. One possible strategy for obtaining desirable proteins is modification of naturally obtained protein based on rational design using molecular modeling and the following site-directed mutagenesis to make actual mutant proteins. However, these trials often resulted in vain, mainly owing to lack of the precise understanding among three dimensional structure, folding, and mechanism.

On the other hand, an alternative approach has been adopted in the past decade, which mimics the natural evolution process consisting of repetitive cycles of genetic diversification, gene expression, and selection of mutants with an improved property in laboratory, to search for desired protein for each use effectively. To explore a vast diversity in an acceptable time, both high-throughput protein expression systems and high-throughput screening scheme are necessary.

Several cell-based peptide display techniques such as phage display *(1–3)*, peptide-on-plasmids *(4)*, bacterial cell-surface display *(5)*, and yeast cell-surface display *(6–8)*, have been developed and used practically. However, they sometimes cannot be applied to toxic proteins owing to the necessity of cell-growth, and their library size is physically constrained by transformation efficiency and the use of agar plates. In vitro or cell-free protein expression system is particularly attractive for high-throughput applications because in vivo manipulations, which are somewhat laborious and time-consuming, can be bypassed. In addition, the expression of proteins that are toxic or harmful to cells is possible *(9,10)*. There are a few technologies for laboratory molecular evolution based on in vitro expression, such as ribosome display *(11–13)*, mRNA–protein fusion *(14,15)*, and in vitro compartmentalization *(16,17)*, and single-molecule PCR-linked in vitro expression (SIMPLEX), which has been developed by our group and will be described here in detail.

Because PCR products can be used as DNA template in the in vitro expression system *(18)*, construction of a protein library would be possible by sequentially combining parallel amplifications of single gene variants and in vitro expression of the outcome PCR products *(19–23)*. A

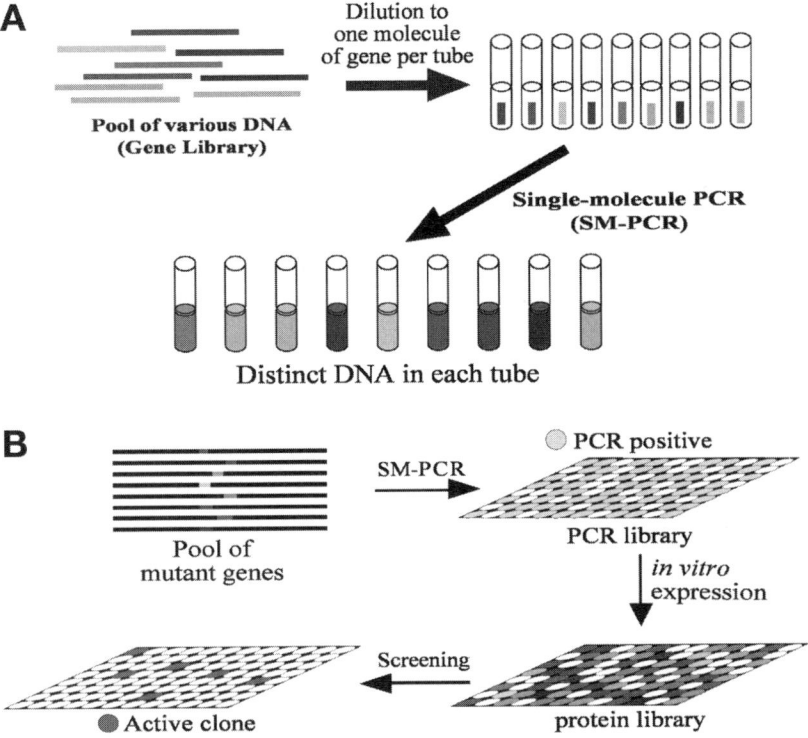

Fig. 1. (**A**) Schematic diagram illustrating the single-molecule PCR (SM-PCR). A gene pool is diluted and single molecule DNAs are dispensed into PCR tubes, and amplified by PCR subsequently. (**B**) Schematic diagram showing a process of the single-molecule PCR-linked in vitro expression (SIMPLEX). A PCR library is generated by parallel SM-PCRs, and subsequently transformed to a protein library by means of in vitro expression. The protein library is, then, subjected to screening.

gene pool is extensively diluted, and each variant gene (single molecule) is dispensed into a well on a PCR plate and subsequently amplified by PCR yielding a one-well-one-gene PCR library (**Fig. 1A**). The PCR amplification of a single-molecule DNA template is called single-molecule PCR (SM-PCR). Thereafter, a protein library is generated by parallel in vitro expression (coupled transcription/translation) of each PCR product from the PCR library. The whole process is called SIMPLEX (**Fig. 1B**). The protein library constructed by means of SIMPLEX is a one-well-one-protein library, so clones in the library can be analyzed individually

without any separation process. Amounts of proteins in a library constructed by SIMPLEX were experimentally confirmed to be highly uni-
form *(21)*. It has been used successfully for high-throughput construction
and screening of mutant libraries of lipase *(24,25)*, manganese peroxidase *(26)*, and scFv *(21,27)*. As all SM-PCRs and in vitro expression
reactions are carried out in parallel on plates, a multichannel (at least 96
channels) dispenser is required for speedy dispensing of reaction mixtures.

## 2. Materials

### 2.1. DNA and Oligonucleotide Primers

1. Gene variants that are placed under control of T7 promoter.
2. A pair of oligonucleotide primers that anneal upstream of T7 promoter
   and downstream of T7 terminator, both should have an identical random
   DNA tag (20–30 bp) at 5'-end (*see* **Notes 1** and **2**).
3. Single primer (or homoprimer): an oligonucleotide primer that has the
   same sequence with the random DNA sequence tag of both ends in a target DNA in **item 2**.

### 2.2. Molecular Biology Kits and Reagents

1. *TaKaRa ExTaq*™ DNA polymerase and its working buffer (Takara Shuzo,
   Kyoto, Japan).
2. *Pfu* Turbo™ DNA polymerase and its working buffer (Stratagene, La
   Jolla, CA).
3. NTPs mix (2.5 m*M* each) for PCR grade.
4. Agarose for electrophoresis grade.
5. Blue Dextran 2000 (GE Healthcare Life Sciences).
6. GeneElute Agarose Spin Columns (Sigma, St. Louis, MO).
7. T7 RNA polymerase (in-house prepared or purchased from suppliers).
8. *Escherichia coli* S30 extract (in-house prepared or purchased from suppliers such as Roche-Diagnostics and Promega).
9. 38 m*M* ATP (GE Healthcare Life Sciences).
10. 500 m*M* GTP (GE Healthcare Life Sciences).
11. 500 m*M* CTP (Sigma).
12. 500 m*M* UTP (Sigma).

13. 2.7 mg/mL Folinic acid (Sigma).
14. 0.1 mg/mL Rifampicin (Sigma).
15. 1.24 *M* Creatine phosphate dipotassium salt (Calibiochem, Darmstadt, Germany).
16. 20 mg/mL Creatine kinase (Boehringer Mannheim).
17. 17.4 mg/mL *E. coli* tRNA (Boehringer Mannheim).
18. 50 m*M* each of 20 amino acid mixture (Wako Pure Chemical Industries, Japan).

## 2.3. Solutions (see Note 3)

1. Phenol/chloroform extraction solution (commonly used for DNA purification. Mixture of phenol equilibriumed with 0.1 *M* Tris-HCl, pH 8.0, chloroform, isoamylalcol [25:24:1]).
2. 3 *M* Sodium acetate, pH 7.0.
3. 95% Ethanol.
4. 70% Ethanol.
5. TE buffer: 10 m*M* Tris-HCl, pH 8.0, 1 m*M* EDTA.
6. 1X TAE buffer: 40 m*M* Tris-acetate pH 8.0, 1 m*M* EDTA.
7. Ethidium bromide 10 mg/mL.
8. 2.2 *M* Tris-acetate, pH 7.4.
9. CK buffer: 20 m*M* Tris-acetate,100 m*M* KCl.
10. 1.4 *M* Ammonium acetate (Wako Pure Chemical Industries), 0.2 *M* magnesium acetate (Wako Pure Chemical Industries).
11. 5.0 *M* Potassium acetate (Wako Pure Chemical Industries).
12. 43.6% w/v Polyethylene glycol 6000 (Wako Pure Chemical Industries).

## 2.4. Devices

1. Spectrophotometer.
2. PCR system (e.g., GeneAmp® PCR system 9700, Applied Biosystems).
3. GeneAmp PCR system 9700 with double 384-well blocks (Applied Biosystems or equivalent devices).
4. Thin wall PCR tubes (Applied Biosystems).
5. 96-Well, round-bottom polystyrene plate.
6. 384-Well PCR plates (Applied Biosystems).
7. Seal for 384-well PCR plates (Applied Biosystems).
8. 15-mL Centrifuge tube.

9. 50-mL Centrifuge tube.
10. Multimicropipet with 8 or 12 channels.
11. Agarose gel electrophoresis apparatus.
12. 96-Head automated dispenser (if available).

## 3. Methods

### 3.1. Preparation of DNA Templates for SM-PCR (see Note 3)

A pool of gene variants can be created by several strategies, such as oligonucleotide-mediated mutagenesis, random mutagenesis, and DNA shuffling. In any case, to express gene variants directly from a DNA fragment in vitro, some special DNA sequences adjacent to an open reading frame are required. Basically, the DNA fragment should contain a promoter, a ribosome-binding site, a target gene, and a terminator sequence consecutively (**Fig. 2**). We use the *E. coli* S30 extract-coupled transcription/translation under the control of T7 promoter for the in vitro expression. Therefore, template for SM-PCR consists of T7 promoter, a ribosome-binding site, a gene variant, and a downstream T7 terminator sequence consecutively. Such a construct is obtainable by cloning gene variants into a typical T7 expression vector, such as pRSET or pET, or using PCR assembly. In addition, a random DNA sequence (~20–30 bp) should be added to the both ends of the template by PCR to be annealing sites for a corresponding single primer in the following SM-PCR. The use of a single primer (homoprimer), which requires the homo-tailed target sequence, can prevent accumulation of the primer-dimer during even more than 50 cycles of amplification *(20,28)*. This is because the primer dimer from a single primers forms a pan-handle structure, a result of self annealing of head to tail, which can not accept new primers, resulting in the prevention of further amplification of the primer dimers.

Here, we will start from adding the carefully designed random DNA tag to both ends of the construct (plasmid or PCR-assembled fragment) that already has all sequences necessary for in vitro-coupled transcription/translation in place.

1. Add a random DNA tag to both ends of DNA templates by PCR, using a forward primer that anneals upstream of T7 promoter and a reverse primer

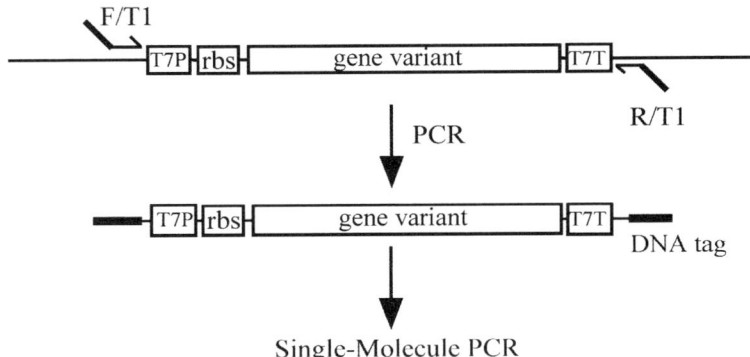

Fig. 2. Illustration of single-molecule PCR (SM-PCR) template preparation. Gene variant is placed under control of T7 promoter (T7P) by cloning or PCR assembly, resulting in a plasmid or a DNA fragment capable of being expressed in vitro. Then, a random DNA tag is appended to both ends of the construct by PCR using a pair of primers that have an identical 5'-end.

that anneals downstream of T7 terminator. Both primers have an identical DNA tag (20–27 bp) at 5'-end. Set up a PCR mixture as follows:

| | |
|---|---|
| DNA templates | 10–50 ng |
| 10X ExTaq Buffer | 10 µL |
| dNTPs (2.5 m$M$ each) | 10 µL |
| ExTaq DNA polymerase | 2.5 U |
| Forward primer | 1 µ$M$ |
| (e.g.,CCATTCATTAATGCACATAACTATTCCATCTCGATCCCGCGAAATTAATACG) | |
| Reverse primer | 1 µ$M$ |
| (e.g., CCATTCATTAATGCACATAACTATTCCTCCGGATATAGTTCCTCCTTTCAG) | |
| Sterile water adjust to | 100 µL |

2. Perform PCR thermal cycling: preheat at 94°C for 5 min; 25 cycles of 94°C for 15 s, 53°C* for 15 s and 72°C for 1 min 20 s; additional extension at 72°C for 7 min. Annealing temperature (*) has to be adjusted according to the $T_m$ of each primer.

3. Analyze 10–20 mL of the PCR products by electrophoresis on 1% agarose gel containing 0.5 mg/mL ethidium bromide. Purify DNA fragments of appropriate size using the GeneElute Agarose Spin Columns.

4. Extract DNA once with phenol/chloroform and precipitate it by adding 0.1 vol of 3 $M$ sodium acetate and 2 vol of cold ethanol.

5. After centrifugation at maximum speed for 10 min, remove ethanol, wash with 70% ethanol, and dry the DNA pellet up by vacuum centrifugation for 5–10 min, until all residual ethanol is evaporated.

6. Dissolve DNA pellet with 20 µL of TE buffer.

7. Use 2 µL of DNA solution to measure its absorbance at 260 nm.

8. Calculate concentration of the DNA solution (1 OD = 50 µg/mL) (*see* **Note 4**).

### 3.2. PCR Library Construction SM-PCR (see Note 5)

To spread DNA molecules on PCR plates at one DNA molecule per well on average, thus prepared SM-PCR templates should be precisely diluted according to the estimation, and a calculated volume of solution containing one molecule should be dispensed to each well of PCR plates. A general equation to calculate the dilution rate is given in **Note 6** at the end of this protocol. Each template is then separately amplified using a high-fidelity DNA polymerase (e.g., *Pfu* Turbo DNA polymerase) for a number of cycles until enough amount of a PCR product is achieved. Distribution of DNA template per well when using the stochastic dilution is in accordance with the Poisson distribution *(29)*, i.e., even if the template concentration is completely precise and the efficiency of amplification from one molecule is 100%, only 63% of all reactions is calculated to have a band, and only 37% of the trials have one molecule origin, and the other 26% of reactions contain two or more molecules of original template. Nevertheless, when a reaction that contains multiple variants gives a positive and promising result in screening, the PCR product can be diluted and used as templates for another round of screening to get actual individual positive genes.

In addition, the SM-PCR can be done with multiple templates. Even if five molecules exist in the same tube, each one can be amplified at almost the same efficiency. Therefore, variation is maintained after more than 60 cycles of PCR. If probability of positive clone is estimated to be relatively low, two-step SIMPLEX strategy, that begin with first screening of positive wells containing at least one positive and many negative genes, by multiple (five molecules per well ore more) PCR, in vitro expression, and screening, then followed by the SM-PCR from the pooled genes, in vitro expression and screening again. The strategy can increase the throughput of the SIMPLEX technology very easily.

1. Convert the concentration of SM-PCR templates (homo-tailed DNA fragments in **Subheading 3.1.**) from the unit of microgram per milliliter to molecule per milliliter.

   1 kb ds DNA = $6.6 \times 10^5$ D

   1 mol has $6.02 \times 10^{23}$ molecules

   conc. in molecules/mL = [(conc. in μg/mL) $\times 6.02 \times 10^{23} \times 10^3$] / [(no. of bp in a template) $\times 6.6 \times 10^5 \times 10^6$]

   If there is variation in the sizes of SM-PCR templates, use the average value as representative.

2. Dilute the SM-PCR templates by 0.01% blue dextran (for preventing non-specific absorption onto the wall of plastic tubes) in TE buffer to the final concentration of 1 molecule/μL.

3. Turn on the GeneAmp PCR system 9700 with double 384-well blocks (PE Applied Biosystems) and set thermal cycle as follows:

   Preheat: 94°C 5 min, and 65 cycles of:

   | | | |
   |---|---|---|
   | Denature | 96°C | 5 s |
   | Annealing | $T_m$°C | 5 s ($T_m$ is melting temperature of the single primer calculated by an appropriate software. We generally use Tm determinator on web) |
   | Extension | 72°C | × min (extension time depends on size of the target DNA and the polymerase used. For 1-kbp fragment amplification by the *Pfu* turbo DNA polymerase, 80 s was actually used.) |

   Add 72°C for 7 min to complete all extension, and put on hold at 4°C.

4. Set up PCR mixture, on ice, for each reaction as follows (order of mixing is important) (*see* **Notes 7** and **8**):

   | | 1 reaction | 1000 reactions |
   |---|---|---|
   | Sterile water | 3.20 μL | 3.20 mL |
   | 10X Cloned *Pfu* Buffer | 0.50 μL | 0.50 mL |
   | dNTPs (2.5 m*M* each) | 0.25 μL | 0.25 mL |
   | Single primer (100 μ*M*) | 0.05 μL | 50.0 μL |
   | (e.g., CCATTCATTAATGCACATAACTATTCC) | | |
   | SM-PCR templates | 1.00 μL | 1.00 mL |
   | *Pfu Turbo*™ DNA polymerase | 0.10 μL | 100.0 μL |

   To carry out 768 SM-PCR reactions at a time on two 384-well PCR plates, mixture for 900–1000 reactions is needed depending on quality of the dispenser used. For this protocol, we use a 96-head dispenser for dispensing of reaction mixtures.

5. Put a 96-well, round-bottom (Falcon) polystyrene plate on ice and dispense 50 µL (for 8 of 5 µL SM-PCR reactions with some excess mixture) of SM-PCR reaction mixture (**step 4**) into each well of the plate. To speed up the process, multichannel micropipet is preferable. This plate serves as SM-PCR reaction reservoir for dispensing the reaction mixture into two 384-well PCR plates.

6. Dispense 5 µL of SM-PCR reaction mixture (**step 5**) to each well of two 384-well PCR plates using an automated 96-channel dispenser.

7. Cover the PCR plates with seals.

8. Spin down SM-PCR reaction mixture.

9. Put the PCR plates onto the thermocycler and start the thermal cycling.

10. After the programed thermocycling, pick several SM-PCR reactions up randomly to check amplification of this batch by agarose gel electrophoresis. Sixty-six to seventy percent of SM-PCR reactions should give a band of expected size. If all SM-PCRs give the band, it is likely that self-contamination of once amplified fragment has occurred, and more than one molecule of templates may have been present in all SM-PCR reactions. If the efficiency is less than 50%, estimation of the amount of the SM-PCR template or the dilution process may not have been accurate.

## 3.3. Protein Library Generation by In Vitro-Coupled Transcription/Translation

Prior to performing in vitro-coupled transcription/translation to generate a protein library from the PCR library, expression efficiency of the target proteins should be examined preliminary (*see* **Note 9**). The obtained data is used to determine the volume of in vitro expression reaction that is related to amount of protein required for the following screening. In addition, amount of SM-PCR product to use as template for the reaction must be optimized. Here, we present a protocol for 20 µL in vitro-coupled transcription/translation reactions, using 2 µL of SM-PCR products as templates, to generate a protein library from a SM-PCR library on two 384-well plates.

1. Chill two of 384-well, flat bottom polystyrene plate (Nunc) and a 96-well, round-bottom (Falcon) polystyrene plate on ice.

2. Prepare a mixture of low molecular weight components for the in vitro-coupled transcription/translation by using *E. coli* S30 extract with T7 RNA polymerase system in a 15-mL tube as follows:

| Low molecular weight mixture (LM) | Volume | Final concentration |
|---|---|---|
| 2.2 *M* Tris-acetate (pH 7.4) | 800 µL | 56.4 m*M* |
| 38 m*M* ATP | 1.0 mL | 1.22 m*M* |
| 88.4 m*M* each GTP, UTP, CTP mixture | 300 µL | 0.85 m*M* each |
| 1.24 *M* Creatine phosphate dipotassium salt | 1.0 mL | 40.0 m*M* |
| 50 m*M* each 20 amino acid mix | 200 µL | 0.32 m*M* each |
| 43.6% w/v PEG 6000 | 2.90 mL | 4.0% w/v |
| 2.7 mg/mL Folinic acid | 400 µL | 34.6 µg/mL |
| 1.4 *M* Ammonium acetate | 800 µL | 35.9 m*M* |
| 17.4 mg/mL tRNA | 300 µL | 0.17 mg/mL |
| Sterile water | 100 µL | |
| Total | 7.80 mL | |

Vortex the mixture of 20 amino acids well before pipetting. Always prepare the mixture on ice. The LM can be stored at –4°C until use.

3. For 768 in vitro-coupled transcription/translation reactions (384 × 2), reaction mixture for 1000 reactions is prepared in a 50-mL tube as follows:

| Reaction mixture | Volume | Final concentration |
|---|---|---|
| LM | 5.0 mL | |
| 20 mg/mL Creatine kinase (in CK buffer) | 150 µL | 0.15 mg/mL |
| 2.3 mg/mL T7 RNA Polymerase | 86.7 µL | 10.0 µg/mL |
| 0.1 mg/mL Rifampicin | 2.0 mL | 0.01 mg/mL |
| 0.2 *M* Magnesium acetate | 670 µL | 6.7 m*M* |
| 5 *M* Potassium acetate | 480 µL | 120 m*M* |
| *E. coli* S30 extract (*see* **Note 10**) | 5.67 mL | |
| Sterile water | 3.94 mL | |

Always keep reaction mixture on ice and mix well every step by turning upside down, do not vortex.

4. Dispense 170 µL of the reaction mixture (**step 3**) into each well of the chilled, 96-well, round-bottom plate using multichannel micropipet. This plate serves as a reservoir to dispense the reaction mixture into two of 384-well PCR plates.

5. Dispense 18 µL of the reaction mixture (**step 4**) to each well of two chilled, 384-well polystyrene plates using the automatic 96-channel dispenser.
6. Dispense 2 µL of SM-PCR products from the PCR library to the 384-well, polystyrene plates containing the in vitro-coupled transcription/translation reaction mixture (*5*) and mix well using the automatic 96-channel dispenser. Now, the position of an SM-PCR product on the SM-PCR plate is identical to the position of its encoded protein on the in vitro expression plates (genotype-phenotype linkage is kept in form of position on plate).
7. Seal the reaction plates, spin down, and incubate them for 1 h 30 min at an appropriate temperature depending on target proteins; for example, 30°C for the expression of scFv.
2. 8. Incubation time and temperature for in vitro-coupled transcription/translation is adjustable depending on target proteins.
9. After incubation, the protein library is ready for screening.

## 4. Notes

1. The random DNA tag must be carefully designed to prevent self-annealing and unspecific amplification of easily found contaminants that could inhibit or interfere the SM-PCR. To facilitate the design of an appropriate terminal sequence for SM-PCR to avoid primer dimer formation and misanealing, the criteria of which should be mush more restricted than normal PCR, a software named "Homoprimer" run on Mac OS9 would be available from the corresponding author.
2. Always use oligonucletide primers of high quality and purity (HPLC grade recommended.
3. General solutions for molecular biology should be prepared as described in "Molecular Cloning" (*30*).
4. Store SM-PCR template as concentrated DNA, and freshly dilute it before use. Diluted DNA (1 molecule/µL) tends to be degraded over storage.
5. In the SM-PCR where number of DNA template is presumably only one molecule and several cycles of thermocycling are performed, cross- or carry-over contamination is a major concern, and sometimes occurs indeed. The followings are guide lines to avoid this trouble:
   a. Wear gloves while working for SIMPLEX, and change gloves as often as possible.
   b. Aliquot an appropriate amount of DNA polymerase and its working buffer, and all other reagents into several tubes for storage, and use each tube only once.
   c. All reagents and buffers should be prepared freshly.

d. Use different area (rooms) for SM-PCR template preparation, SM-PCR template dilution, SM-PCR setup, and expression.

e. Use pipet tips with filter and separated micropipet sets for specific area.

f. When dispensing solution with an auto-pipetter, do not repeat sucking in/pushing out.

g. Do not use the same primer for different libraries.

6. A general equation to calculate the precise dilution rate to give rise to n numbers of the targeted DNA molecule (template) in one final PCR tube (i.e., n-molecule PCR) is given by the following equation:

$$D = \frac{CV}{N_{bp}n} \times \frac{6.02 \times 10^{14}}{660}$$

where $V$ is the volume of the diluted DNA solution, which is to be put into a PCR tube (µL), $C$ is the concentration of the DNA fragment in the original DNA solution (µg/mL), $N_{bp}$ is the number of basepairs of the targeted DNA fragment, and $n$ is the number of the targeted DNA molecule in each PCR tube. For single molecule per tube, n = 1, $D$ should from $10^{10}$ to $10^{12}$.

7. Always set up reactions, both SM-PCR and in vitro-coupled transcription/translation, on ice to prevent autonomous start of reactions.

8. Quality of PCR library and protein library is highly related to speed of reaction setup. The faster the better.

9. Efficiency of the in vitro-coupled transcription/translation dramatically depends on quality of extracts. For in-house prepared *E. coli* S30 extracts where quality might be different in each batch, quality control of the batches is very important. It is the best to use *E. coli* S30 extract from one batch for construction of a library.

10. To express protein with disulfide-bridge, use *E. coli* S30 extract without dithiothreitol, or add appropriate amount of glutathione/oxidized glutathione to adjust the redox condition for the formation of proper disulfide bonds.

## References

1. Mao, S., Gao, C., Lo, C. L., Wirshing, P., Wong, C., and Janda, K. D. (1999) Phage-display library selection of high-affinity human single-chain antibodies to tumor-associated carbohydrate antigens sialyl Lewisx and Lewisx. *Proc. Natl. Acad. Sci. USA* **96**, 6953–6958.

2. Zavala, A. G., Lancaster, T., Groopman, J. D., Stickland, P. T., and Chandrasegaran, S. (2000) Phage display of ScFv peptides recognizing the thymidine(6-4)thymidine photoproduct. *Nucleic Acids Res.* **28**, e24.

3. Rodi, D. J. and Makowski, L. (1999) Phage-display technology- finding a needle in a vast molecular haystack. *Curr. Opin. Biotechnol.* **10**, 87–93.
4. Cull, M. G., Miller, J. F., and Schatz, P. J. (1992) Screening for receptor ligands using large libraries of peptides linked to the C terminus of the *lac* repressor. *Proc. Natl. Acad. Sci. USA* **89**, 1865–1869.
5. Georgiou, G., Stathopoulos, C., Daugherty, P. S., Nayak, A. R., Iverson, B. L. and Ill, R. C. (1996) Display of heterologous proteins on the surface of microorganisms: from the screening to live recombinant vaccines. *Nat. Biotechnol.* **15**, 29–33.
6. Boder, E. T. and Wittrup, K. D. (1997) Yeast surface display for screening combinatorial polypeptide libraries. *Nat. Biotechnol.* **15**, 553–557.
7. eke, M. C., Cho, B. K., Boder, E. T., Kranz, D. M., and Wittrup, K. D. (1997) Isolation of anti-T cell receptor scFv mutants by yeast surface display. *Protein Eng.* **10**, 1303–1310.
8. Cho, B. K., Kieke, M. C., Boder, E. T., Wittrup, K. D., and Kranz, D. M. (1998) A yeast surface display system for the discovery of ligands that trigger cell activation. *J. Immunol. Methods* **220**, 179–188.
9. Nakano, H. and Yamane, T. (1998) Cell-free protein synthesis systems. *Biotechnol. Adv.* **16**, 367–384.
10. Jermutus, L., Ryabova, L. A., and Plückthun, A. (1998) Recent advances in producing and selecting functional proteins by using cell-free translation. *Curr. Opin. Biotechnol.* **9**, 534–548.
11. Mattheakis L. C., Bhatt R. R., and Dower W. J. (1994) An *in vitro* polysome display system for identifying ligands from very large peptide libraries. *Proc. Natl. Acad. Sci. USA* **91**, 9022–9026.
12. Hanes, J. and Plückthun, A. (1997) *In vitro* selection and evolution of functional proteins by using ribosome display. *Proc. Natl. Acad. Sci. USA* **94**, 4937–4942.
13. Hanes, J., Schaffitzel, C., Knappik, A., and Plückthun A. (2000) Picomolar affinity antibodies from a fully synthetic naïve library selected and evolved by ribosome display. *Nature Biotechnol.* **18**, 1287–1292.
14. Nemoto, N., Miyamoto-Sato, E., Hushimi, Y., and Yanagawa, H. (1997) *In vitro* virus: Bonding of mRNA bearing puromycin at the 3'-terminal end to the C-terminal end of its encoded protein on the ribosome *in vitro*. *FEBS Lett.* **414**, 405–408.
15. Roberts, R. W. and Szostak, J. W. (1997) RNA-peptide fusions for the *in vitro* selection of peptides and proteins. *Proc. Natl. Acad. Sci. USA* **94**, 12,297–12,302.
16. Tawfik, D. S. and Griffiths, A. D. (1998) Man-made cell-like compartments for molecular evolution. *Nature Biotechnol.* **16**, 652–656.

17. Griffiths, A. D. and Tawfik, D. S. (2003) Directed evolution of an extremely fast phosphotriesterase by in vitro compartmentalization. *EMBO* **22,** 24–35.

18. Martemyanov, K. A., Spirin, A. S., and Gudkov, A. T. (1997) Direct expression of PCR products in a cell-free transcription/translation system: synthesis of antibacterial peptide cecropin. *FEBS Lett.* **414,** 268–270.

19. Ohuchi, S., Nakano, H., and Yamane, T. (1998) *In vitro* method for the generation of protein libraries using PCR amplification of a single DNA molecule and coupled transcription/translation. *Nucleic Acids Res.* **26,** 4339–4346.

20. Nakano, H., Kobayashi, K., Ohuchi, S., Sekiguchi, S., and Yamane, T. (2000) Single-step single-molecule PCR of DNA with a homo-priming sequence using a single primer and hot-startable DNA polymerase. *J. Biosci. Bioeng.* **90,** 456–458.

21. Rungpragayphan, S., Kawarasaki, Y., Imaeda, T., Kohda, K., Nakano, H., and Yamane, T. (2002) High-throughput, cloning-independent protein library construction by combining single-molecule DNA amplification with *in vitro* expression. *J. Mol. Biol.* **318,** 395–405.

22. Rungpragayphan, S., Nakano, H., and Yamane, T. (2003) PCR-linked *in vitro* expression: a novel system for high-throughput construction and screening of protein libraries. *FEBS Lett.* **540,** 147–150.

23. Nakano, H., Kawarasaki, Y., and Yamane, T. (2004) Cell-free protein synthesis systems: Increasing their performance and applications. *Adv. Biochem. Engin/Biotechnol.* **90,** 135–149.

24. Koga, Y., Kobayashi, K., Yang, J., Nakano, H., and Yamane, T. (2002) In vitro construction and screening of a *Burkholderia cepacia* lipase library using single-molecule PCR and cell-free protein synthesis. *J. Biosci. Bioeng.* **94,** 84–86.

25. Koga, Y., Kato, K., Nakano, H., and Yamane, T. (2003) Inverting enantioselectivity of Burkholderia cepacia KWI-56 lipase by combinatorial mutation and high-throughput screening using single-molecule PCR and in vitro expression. *J. Mol. Biol.* **331,** 585–592.

26. Miyazaki-Imamura, C., Oohira, K., Kitagawa, R., Nakano, H., Yamane, T., and Takahashi, H. (2003) Improvement of H2O2 stability of manganese peroxidase by combinatorial mutagenesis and high-throughput screening using in vitro expression with protein disulfide isomerase. *Protein Eng.* **16,** 423–428.

27. Rungpragayphan, S., Haba, M., Nakano, H., and Yamane, T. (2004) Rapid screening for affinity-improved scFvs by means of single-molecule-PCR-linked *in vitro* expression. *J. Mol. Catalysis B: Enzymatic* **28,** 223–228.

28. Brownie, J., Shawcross, S., Theaker, J., et al. (1997) The elimination of primer-dimer accumulation in PCR. *Nucleic Acids Res.* **25,** 3235–3241.
29. Stephens, J. C., Rogers, J., and Ruano, G. (1990) Theoretical underpinning of the single-molecule-dilution (SMD) method of direct haplotype resolution. *Am. J. Hum. Genet.* **46,** 1149–1155.
30. Sambrook, J. and Russell, D. W. (2001) *Molecular Cloning: A Laboratory annual, 3rd ed.* Cold Spring Harbor Laboratory Press, NY.

# 5

# Methods for High-Throughput Materialization of Genetic Information Based on Wheat Germ Cell-Free Expression System

Tatsuya Sawasaki, Ryo Morishita, Mudeppa D. Gouda, and Yaeta Endo

## Summary

Among the cell-free protein synthesis systems, the wheat germ-based translation system has significant advantages for the high-throughput production of eukaryotic multidomain proteins in folded state. Here, we describe protocols for this cell-free expression system.

**Key Words:** Cell-free protein synthesis; wheat germ extract; 5'- and 3'-UTRs; PCR; using split-primers; transcription and translation; purification of products.

## 1. Introduction

This chapter describes in detail the methods for high-throughput protein production based on the cell-free system prepared from eukaryotic wheat embryos. The methods are divided into four steps as follows: (1) preparation of the highly efficient extract from wheat embryos; (2) generation of DNA template for transcription of the desired open reading frame (ORF) with or without cloning; (3) sequential transcription-trans-

From: *Methods in Molecular Biology,* vol. 375,
*In Vitro Transcription and Translation Protocols, Second Edition*
Edited by: G. Grandi © Humana Press Inc., Totowa, NJ

lation by using PCR generated DNA or highly purified plasmid carrying ORF based on the bilayer reaction method, and (4) purification of synthesized protein by on-column digestion. Purification of in vitro-transcribed products is successfully carried out by synthesizing the proteins as fused forms and exploiting the fused tag. Finally, the tag is removed proteolytically. This platform is already in use for functional and structural analyses of gene products *(1,2)*.

## 2. Materials

1. Wheat seeds.
2. Roter Speed Mill (model pulverisette 14, Fritsh, Germany).
3. Sieve (710- to 850-μm mesh).
4. Cyclohexane and carbon tetrachloride.
5. Nonidet P-40.
6. Milli-Q water, freshly prepared.
7. Bronson model 2210 sonicator (Yamato, Japan).
8. Extraction buffer: 40 m$M$ HEPES-KOH, pH 7.6, 100 m$M$ potassium acetate, 5 m$M$ magnesium acetate, 2 m$M$ calcium chloride, 4 m$M$ DTT, and 0.3 m$M$ of each of the 20 amino acids.
9. Sephadex G-25 (fine) (Amersham Biosciences).
10. pEU protein expression vector.
11. cDNA of protein to be synthesized.
12. Oligonucleotide primers.
13. PCR thermo-cycler MP (Takara, Otsu, Japan).
14. ExTaq DNA polymerase (Takara).
15. 5X Transcription buffer (TB): 400 m$M$ HEPES-KOH, pH 7.8, 80 m$M$ magnesium acetate, 10 m$M$ spermidine, and 50 m$M$ DTT.
16. Nucleotide tri-phosphates (NTPs) mix: a solution containing ATP, GTP, CTP, and UTP (25 m$M$ each).
17. SP6 RNA polymerase and RNasin (80 U/μL, Promega, Madison, WI).
18. Microcon (YM-50; Millipore, Bedford, MA).
19. Amicon Ultra-15 (10 K, Millipore, cat. no., UFC901024).
20. 2 mg/mL (20 mg/mL for large-scale production) creatine kinase.
21. Translational substrate buffer (TSB): 30 m$M$ HEPES-KOH, pH 7.8, 100 m$M$ potassium acetate, 2.7 m$M$ magnesium acetate, 0.4 m$M$ spermidine, 2.5 m$M$ DTT, 0.3 m$M$ amino acid mix, 1.2 m$M$ ATP, 0.25 m$M$ GTP, and 16 m$M$ creatine phosphate.

22. 24-Well microplate (Whatman Inc., Clifton, NJ) for small-scale production or 6-well microplate (Whatman Inc.) for large-scale production.
23. 1X PBS: 140 m$M$ NaCl, 2.7 m$M$ KCl, 10.1 m$M$ Na$_2$PO$_4$, 1.8 m$M$ KH$_2$PO$_4$.
24. Poly-Prep Chromatography Column (Bio-Rad, cat. no. 731-1550).
25. Glutathione Sepharose 4B (Amersham Biosciences).
26. PreScission Protease (2 U/μL, Amersham Biosciences).

## 3. Methods

### 3.1. Preparation of Wheat Embryos Extract

Isolation of embryos and preparation of extract were carried out as reported previously *(3)*. The main aim is to remove contaminating endosperms from embryos by washing (**Subheading 3.1.1., steps 5** and **6**).

#### 3.1.1. Isolation of Wheat Embryo

1. Grind the wheat seeds in a mill (Roter Speed Mill).
2. Sieve through a 710- to 850-μm mesh.
3. Select the intact embryos by solvent flotation using cyclohexane and carbon tetrachloride as solvents (1:2.5 v/v).
4. Dry overnight in a fume hood.
5. Wash three times with 10 vol of sterile water under vigorous stirring.
6. Sonicate for 3 min in a 0.5% Nonidet P-40 solution by using a Bronson model 2210 sonicator.

#### 3.1.2. Preparation of Wheat Embryo Extract

1. Grind 5 g of isolated wheat embryo to a fine powder in liquid nitrogen.
2. Add 5 mL of extraction buffer and vortex the mixture briefly.
3. Centrifuge the embryo lysate (30,000$g$; 30 min) and retain the supernatant.
4. Gel-filtrate the supernatant by using a sephadex G-25 (fine) column, equilibrated with 2 vol of extraction buffer.
5. Replace the ingredients by the gel-filtration as above but using the TSB as equilibration buffer.
6. Collect the void fraction.
7. Concentrate the fraction to approx 1/3 vol using Amicon Ultra-15 (10 K) filter unit according to the manufacturer's instructions at 10°C.
8. Adjust to 240 A260/mL with the TBS.
9. Divide into small aliquots, and store at –80°C until use.

## 3.2. DNA-Template Construction for Transcription

Cap and poly A (pA), found on almost all eukaryotic mRNAs, stimulate translation initiation and stabilize mRNA by their synergistic action. However, for in vitro translation the use of cap and pA can be a major problem *(4)*. This has been effectively overcome by inserting optimized 5'- and 3'-untranslated regions (UTRs) upstream and downstream from the coding regions *(4)*. Optimized 5'- and 3'-UTRs can be fused to the gene of interest in two different ways: (1) by cloning the same gene in a plasmid vector carrying the proper UTRs and then amplifying the gene attached to 5'- and 3'-UTRs from the plasmid, and (2) by fusing the 5'- and 3'-UTRs during the amplification reaction from cDNAs. The optimized elements are assembled into a pSP65-derived vector called pEU **(Fig. 1A,B)**. By using this assembled vector, the desired genes can be expressed efficiently without cap and poly(dA/dT) *(4)*. Cloning of the desired gene into the pEU3b expression system is achieved by using standard methods. After confirming the DNA sequence, the plasmid containing the desired ORF is prepared by either the CsCl method *(5)* or by using small-scale plasmid preparation kits. Prior to mRNA synthesis, recombinant plasmid should be treated with phenol-chloroform to eliminate any RNases activity.

### 3.2.1. Template Preparation From pEU Carrying Optimized 5'- and 3'- UTRs

We have optimized a PCR-based linear template DNA preparation by designing a set of universal primers for pEU vector. The designed primers are SPu (5'- GCGTAGCATTTAGGTGACACT) and AODA2303 (5'- GTCAGACCCCGTAGAAAAGA). By using these primers, the desired DNA template can be easily generated and used directly for the preparation of mRNA without further purification as follows.

1. Prepare 100 µL of PCR reaction mixture as per the manufacturer's instructions (Takara) by mixing 100 pg/µL of plasmid and 200 n*M* of each primer, 200 µ*M* of each dNTP, and 1.25 U of ExTaq DNA polymerase.
2. Set the PCR thermo-cycler (Takara) on 1 min denaturation at 96°C followed by 25 cycles of amplification: 98°C for 10 s, 55°C for 30 s, and 72°C for 5 min depending on the length of the ORF (1 kb per min).

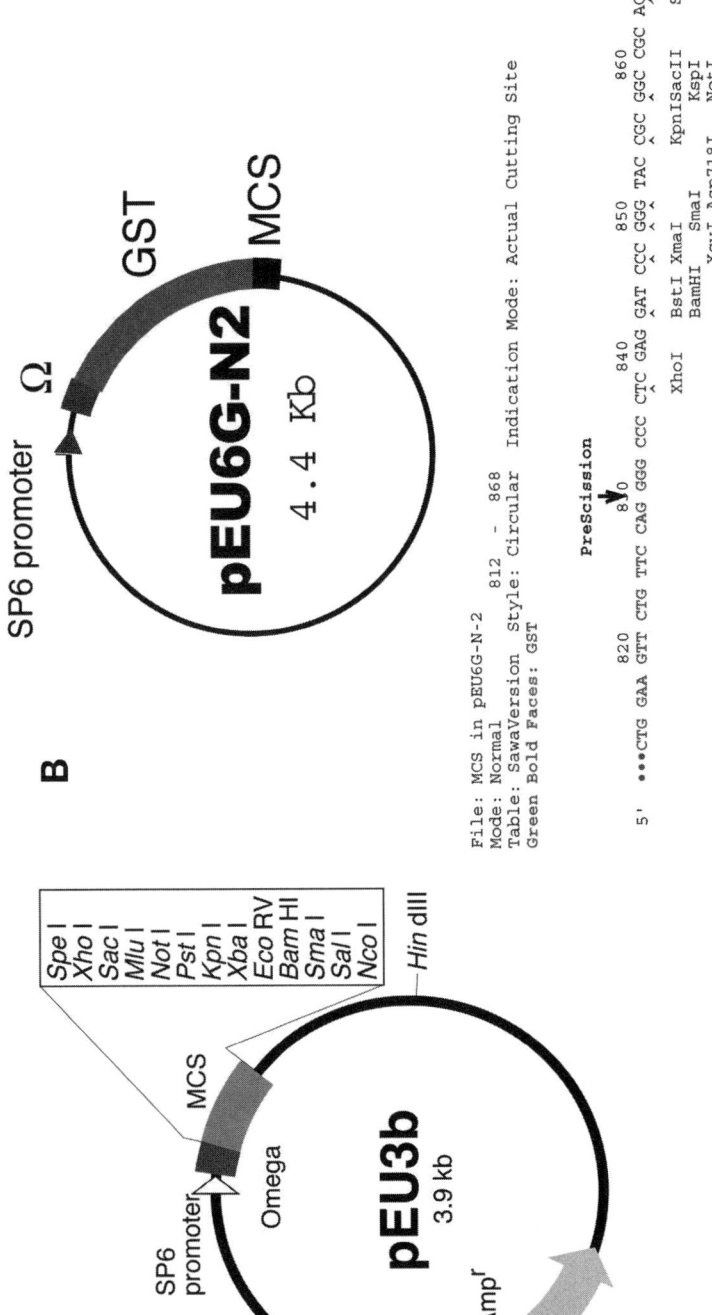

Fig. 1. (A) Schematic diagram of the pEU expression vector. (B) Map and sequence in multicloning site of pEU6G-N2 for GST-tag purification.

3. Concentrate the PCR product to approx 20 μL (~5 times) using Microcon according to the manufacturer's instructions.
4. After analyzing the integrity of DNA template by agarose gel electrophoresis, the template DNA is used for the transcription.

## 3.2.2. Direct Fusion of 5'- and 3-UTRs by Split-Primers PCR Technique

Because cloning of cDNA into an expression vector is one of the most laborious and time-consuming steps of the entire process, direct synthesis of the template by PCR is preferred *(6,7)*. This is achieved by the incorporation of transcriptional and translational start and stop signals into primers used for the amplification of the gene of interest. In this section, we focus on the construction of template DNA starting from cDNA clones by using the split-primers PCR technique *(4)*. The strategy is illustrated in **Fig. 2**.

### 3.2.2.1. DESIGN OF PRIMERS

As shown in **Fig. 2**, split-primers PCR technique makes use of four primers:

1. The target specific primer (primer 3) is designed in such a way that its 3'-end can anneal to the 5'-end of the target gene and its 5'-end has part of the omega sequence *(4)*.
2. Primer 2 has the full-length omega sequence, thus allowing the annealing to the primer 3-derived amplification product and a part of the SP6 promoter sequence.
3. Primer 1 has the remaining part of the SP6 promoter and can anneal to primer 2-derived amplification product.
4. Primer 4 is specific for the gene-carrying vector and can anneal several hundreds of basepairs downstream from the 3'-end of the gene of interest.

Therefore, for each clone of cDNA, primer 3 is the only specific primer and the remaining primers are common for all cDNAs. The sequences of the promers used are reported below:

Primer 1: 5'-GCGTAGC<u>ATTTAGGTGACACT</u> (the underlined sequence is the 5' half of the promoter).

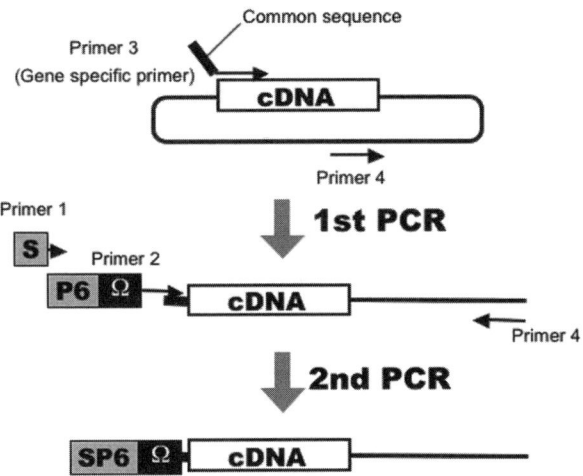

Fig. 2. Direct transcriptional template generation from cDNA library by using the split-primers PCR technique. A schematic representation of the split-primers design for equipping the cDNA sequences with the required UTRs.

Primer 2: 5'-*GGTGACACT*ATAGAAGTATTTTTACAACAATTAC CAACAACAACAACAAACAACAACAACATTACATTTTACATTCT ACAACTA*CCACCCACCACCACCAATG* (underlined is the 3'-half sequence of the promoter and the sequences in italic denote the region annealing to primer 1 and primer 3).

Primer 3: 5'-CCACCCACCACCACCAatgnnnnnnnnnnnnnnnnnnn (the stretch of "ns" indicates the nucleotide sequence which is specific for the gene of interest).

Primer 4: 5'-AGCGTCAGACCCCGTAGAAA (based on the sequence of the vector plasmid described in **refs. 9–11**).

### 3.2.2.2. First PCR

1. Prepare 60 µL of a PCR reaction by mixing 3 ng of plasmid or 3 µL of an overnight culture of *Escherichia coli* carrying the plasmid, of dNTPs (200 µ*M* each ), 1.5 U of ExTaq DNA polymerase, 10 n*M* of primer 3 and primer 4, and the polymerase buffer supplied by the manufacturer.

2. Set the PCR thermo-cycler on 4-min denaturation at 94°C followed by 30 cycles of amplification: 98°C for 10 s, 55°C for 30 s, and 72°C for 4–5 min, depending on the length of the gene (1 kb per min).

### 3.2.2.3. SECOND PCR

To attach the omega and SP6 promoter sequences to the first PCR product, a second 30 μL PCR was carried out by mixing 3 μL of the first PCR product (without any purification), 100 n$M$ of primer 1 and 4, and 1 n$M$ of primer 2 using the same amplification conditions as for the first PCR.

## 3.3. Sequential Transcription-Translation Reactions

Cell-free translation of proteins can be achieved mainly by three different modes, batch mode translation *(3,7)*, bilayer system *(8)*, and continuous-flow cell-free protein synthesis method *(4,9)*. Their basic concepts are described in our previous paper *(10)*. In this section, the bilayer mode of protein synthesis using unpurified mRNA will be explained in detail. **Table 1** summarizes the composition of the mixture, which varies depending on the amount of protein needed.

1. Prepare transcription reaction as indicated in **Table 1** except for wheat embryo extract and creatine kinase.
2. Incubate the reaction mixture at 37°C for 6 h.
3. Add 25 μL (250 μL for large scale [LS]) of wheat embryo extract, and 1 μL of 2 mg/mL creatine kinase (20 mg/mL creatine kinase for LS), and mix well (reaction mixture).
4. Prepare 550 μL (5.5 mL for LS) of TSB in the U-shaped titer plate well (six-well plate for LS).
5. Carefully transfer 50 μL (500 μL for LS) of the reaction mixture to the titer plate well by inserting the pipet tip down to the bottom of the well. Because of the higher density of the **step 3** mix compared with the **step 4** mix, one can clearly see two layers.
6. Place a sealing film and then a cover on the plate to avoid evaporation.
7. Keep the plate in an incubator at 26°C for 18 h without shaking.

**Table 1**
**Reaction Mixture for Transcription and Translation**

| Reagen | Small scale (SS) | Large scale (LS) | Final concentration |
|---|---|---|---|
| Template DNA | $x$ µL | $x$ µL | 1/5 vol (HC PCR)[a] |
| | | | 100 ng/µL (CP)[b] |
| 5X TB | 5 | 50 | 1X |
| 25 m$M$ NTP mix | 2.5 | 25 | 2.5 m$M$ |
| 80 U/µL SP6 RNApolymerase | 0.3 | 3 | 1 U/µL |
| 80 U/µL RNasin | 0.3 | 3 | 1 U/µL |
| Milli-Q | up to 24 | up to 249 | |
| Wheat embryo extract | 25 | 250 | 120 $A_{260}$/mL |
| 2 mg/mL creatine kinase | 1 | – | 40 µg/mL |
| 20 mg/mL creatine kinase | – | 1 | 40 µg/mL |

[a]Highly concentrated PCR product (>50 ng/mL).
[b]High quality circular plasmid.

103

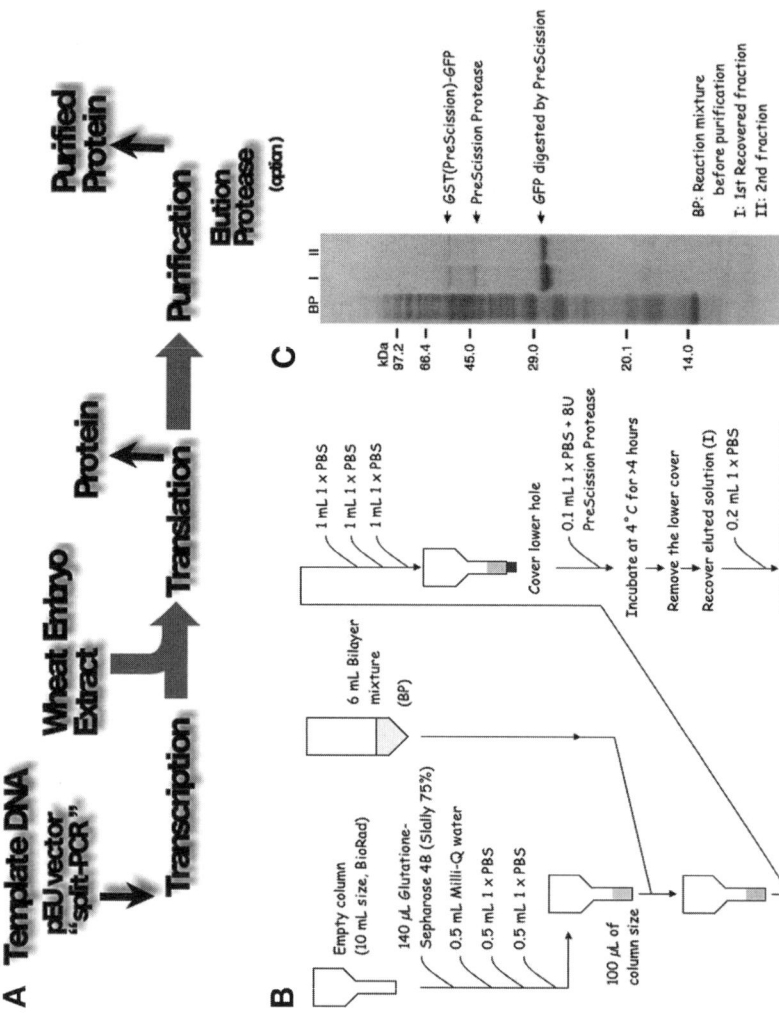

Fig. 3. (A) Flowchart of the cell-free production and purification process based on sequential transcription-translation reaction and purification. (B) Flowchart of protein purification by using GST-tag and on-column digestion with PreScission. (C) An example of purified protein.

## 3.4. Purification of Synthesized Protein by On-Column Digestion

This section describes the methods for (1) isolation of a GST-fused protein as obtained when a pEU-derived transcript is used (**Subheading 3.**), (2) removal of the GST moiety by PreScission protease, and (3) purification of the protein (*see* **Fig. 3B,C**).

1. Prepare the column (10 mL size) by adding 140 µL of glutathione-sepharose 4B resin.
2. Wash the column with 0.5 mL of Milli-Q water followed by 1 mL of 1X PBS.
3. Load 6 mL of the synthesized protein from **Subheading 3.3.** into the column.
4. Wash off the unbound proteins with 3 mL of 1X PBS.
5. Cap the bottom of the column.
6. Treat the column with 8 U PreScission protease in 0.1 mL of 1X PBS and then incubate at 4°C for 4 h.
7. Remove the cap and then elute the target protein without GST tag
8. Recover the target protein with 0.1 mL of 1X PBS one more time.

## 4. Notes

1. A possible procedure for the preparation of proteins can be:
   a. Selection and design of suitable genes from the data bank.
   b. Generation of template for the transcription by PCR, or by cloning into pEU.
   c. Transcription of mRNA.
   d. Translation in the wheat germ cell-free system.
   e. Fine tuning of translation conditions such as ion concentrations and incubation temperature for large-scale protein production.
   f. Purification of the product.

   Because the translation machinery prepared from wheat embryos are efficient and robust, we could in fact succeeded in the robotic automation of the processes from **steps c** through **f**.

# References

1. Sawasaki, T., Hasegawa, Y., Morishita, R., Seki, M., Shinozaki, K., and Endo, Y. (2004) Genome-scale, biochemical annotation method based on the wheat germ cell-free protein synthesis system. *Phytochemistry* **65,** 1549–1555.
2. Vinarov, D. A., Lytle, B. L., Peterson, F. C., Tyler, E. M., Volkman, B. F., and Markley, J. L. (2004) Cell-free protein production and labeling protocol for NMR-based structural proteomics. *Nature Methods* **2,** 1–5.
3. Madin, K., Sawasaki, T., Ogasawara, T., and Endo, Y. (2000) A highly efficient and robust cell-free protein synthesis system prepared from wheat embryos: plants apparently contain a suicide system directed at ribosomes. *Proc. Natl. Acad. Sci. USA* **97,** 559–564.
4. Sawasaki, T., Ogasawara, T, Morishita, R., and Endo, Y. (2002) A cell-free protein synthesis system for high-throughput proteomics. *Proc. Natl. Acad. Sci. USA* **99,** 14,652–14,657.
5. Sambrook, J. and Russell, D. W. (2001) *Molecular Cloning: A Laboratory Manual, 3rd ed.* Cold Spring Harbor Laboratory Press, Cold Spring Harbor, NY.
6. Gurevich, V. V. (1996) Use of bacteriophage RNA polymerase in RNA synthesis. *Methods Enzymol.* **275,** 382–397.
7. Hanes, J. and Pluckthun, A. (1997) In vitro selection and evolution of functional proteins by using ribosome display. *Proc. Natl. Acad. Sci. USA* **94,** 4937–4942.
8. Sawasaki, T., Hasegawa, Y., Tsuchimochi, M., et al. (2002) A bilayer cell-free protein synthesis system for high-throughput screening of gene products. *FEBS Lett.* **514,** 102–105.
9. Spirin, A. S., Baranov, V. I., Ryabova, L. A., Ovodov, S. Y., and Alakhov, Y. B. (1988) A continuous cell-free translation system capable of producing polypeptides in high yield. *Science* **242,** 1162–1164.
10. Endo, Y. and Sawasaki, T. (2004) High-throughput, genome-scale protein production method based on the wheat germ cell-free expression system. *J. Struct. Funct. Genomics* **5,** 45–57.

# 6

# Exogenous Protein Expression in *Xenopus* Oocytes

*Basic Procedures*

## Elena Bossi, Maria Serena Fabbrini, and Aldo Ceriotti

### Summary

The oocytes of the South African clawed frog *Xenopus laevis* have been widely used as a reliable system for the expression and characterization of different types of proteins, including ion channels and membrane receptors. The large size and resilience of these oocytes make them easy to handle and to microinject with different molecules such as natural mRNAs, cRNAs, and antibodies. A variety of methods can then be used to monitor the expression of the proteins encoded by the microinjected mRNA/cRNA, and to perform a functional characterization of the heterologous polypeptides. In this chapter, after describing the equipment required to maintain *X. laevis* in the laboratory and to set up a microinjection system, we provide detailed procedures for oocyte isolation, micropipet and cRNA preparation, and oocyte microinjection. A method for the labeling of oocyte-synthesized proteins and for the immunological detection of the heterologous polypeptides is also described.

**Key Words:** *Xenopus laevis*; oocyte; mRNA; translation; microinjection.

From: *Methods in Molecular Biology, vol. 375,*
*In Vitro Transcription and Translation Protocols, Second Edition*
Edited by: G. Grandi © Humana Press Inc., Totowa, NJ

# 1. Introduction

The *Xenopus laevis* oocyte is a powerful and widely used experimental system that has been extensively exploited for the translation of natural and synthetic mRNAs. In this in vivo expression system the heterologous protein can be subjected to a series of modifications many of which do not occur or occur with reduced efficiency in standard in vitro translation systems. In addition to RNA, other biological materials, such as antibodies *(1)*, can be easily introduced into these large and resilient cells with the help of a relatively simple apparatus. Other distinctive advantages of this expression system are that the expression level can be controlled by varying the amount of injected mRNA *(2)*, and that different proteins can be easily coexpressed in various combinations by simply co-injecting the respective cRNAs *(3)*.

Here, we provide some essential procedures that should allow the reader to exploit *Xenopus* oocytes as a convenient system for the expression of specific proteins. Frog maintenance, oocyte preparation, setup, and use of a microinjection system will be described. A standard procedure that should allow determining whether the desired protein is actually synthesized will be also presented.

# 2. Materials

## 2.1. Frogs

Large *X. laevis* females can be obtained from a number of commercial suppliers: Xenopus I (www.xenopusone.com), Nasco (www.enasco.com), Xenopus express (www.xenopus.com). It is sometime possible to choose between wild, laboratory-conditioned, and laboratory-reared animals. When ordering, it should be specified that the animals should be mature (preferably more than 2 yr old), large or extra large, and obviously female (*see* **Note 1**).

## 2.2 Maintenance of Frogs (see *Notes* 2 and 3)

1. A room in which the temperature is constantly held at 18–22°C, and where a 12 h light/dark cycle is maintained to reduce seasonal variability of oocyte quality. Lamps should simulate sunlight.

2. Glass or plastic water tanks.
3. Plastic gloves.
4. A soft net.
5. Frog food.

## *2.3. Special Items and Equipment*

1. Stereomicroscope: a good stereomicroscope with a lens combination giving a 10-fold magnification and a working distance of about 10 cm will suffice, but a zoom stereomicroscope can be useful to suit personal preferences and will allow to choose the best magnification for each different task (oocyte isolation, selection, micropipet preparation, microinjection).
2. Microinjector: several types of suitable microinjectors are available on the market. We use a microinjection system with direct piston displacement and microprocessor control (Nanoliter injector, WPI, www.wpiinc.com). By setting the DIP switch the injection volume can be changed from 2.3 to 69 nL. This system allows the repeated injection of a calibrated volume by pushing a button on the control box or pressing a footswitch.
3. Micromanipulator: the large size of the oocytes used for microinjection makes the use of a high power manipulator unnecessary. Rather, it is more important to have a good range of movement (around 3 cm) and to be able to move the micropipet quickly from one position to the other. We use the MM-3 micromanipulator (Narishige) with a special adapter for the Nanoliter injector that can be purchased from WPI. A magnetic holding device can be used to fix the micromanipulator on a steel mounting surface.
4. Micropipet puller, capillaries, and micropipet preparation: commercially available micropipet pullers (e.g., PN-3 Horizontal Glass Puller, Narishige) are the best tools to prepare micropipets. The Nanoliter injector uses borosilicate glass capillaries of 1.14 mm OD and 0.5 mm ID, 3.5 in. long (WPI, cat. no. 4878). The high temperature of the pulling process ensures RNase inactivation. The pulled micropipets can be stored for 1 d in a special micropipet storage jar (WPI) or lying on a strip of modeling wax in a 90-mm Petri dish, so that the tips do not get damaged.

   Just before use, the micropipet is placed on the stereomicroscope stage and broken using the walls of a 35-mm sterile Petri dish lid, to obtain an about 10-μm ID tip.

   Prepulled micropipets of 10 μm ID and suitable for use with the Nanoliter injector are also commercially available (WPI, cat. no. TIP10XV119).

5. Light source: although any microscope lamp would be adequate, we recommend the use of a fiber-optic light source. These instruments provide a very good illumination without causing undue warming of the oocytes.

6. Incubator: for most analytical uses, oocytes are cultured at a temperature of 19–20°C. This requires the use of a cooled incubator.

## 2.4. Other Items

1. Watch-maker forceps (e.g., Dumont no. 5).
2. Fluid paraffin.
3. Silicon tubing fitted on a 2-mL syringe.
4. Loose-fitting glass homogenizers: size (0.5–2 mL) will depend on the number of oocytes to be homogenized.
5. Surgery tools: a standard kit of surgical instruments will include two dissecting spring scissors (straight sharp, 10 mm blades), one surgical straight scissors, two Dumont forceps no. 7 (large radius curved shanks), two Iris forceps, one needle holder, suture needles (17.4 mm half circle round bodied), silk, and Dar-Vin absorbable synthetic polyglycolic acid suture threads 4/0 (Ergon Suturamed, Ciba Geigy Group).

## 2.5. Media and Solutions

### 2.5.1. Oocyte Culture Media

1. ND96 Ø $Ca^{2+}$ High Salt Stock (20X): 112.205 g NaCl, 2.982 g KCl, 4.066 g $MgCl_2$, and 23.830 g HEPES free acid. Dissolve in 800 mL of distilled sterile water and adjust the pH to 7.6 with 1 $M$ NaOH. Make up to 1 L and sterilize by autoclaving or filtering. Store as 50-mL aliquots at 4°C for up to 1 mo or at –20°C for up to 6 mo.

2. $CaCl_2$ stock (1 $M$): dissolve 36.755 g $CaCl_2 \cdot 2H_2O$ in 250 mL of distilled sterile water, store at 4°C for up to 1 mo.

3. Piruvate stock (1 $M$): dissolve 27.51 g of Na pyruvate in 250 mL of distilled sterile water. Store as 2.5-mL aliquots in sterile tubes at –20°C for up to 6 mo.

4. Gentamycin stock 250 mg/mL: dissolve 5 g of gentamycin sulfate in 20 mL of distilled sterile water, store as 250-µL aliquots in 1.5 mL sterile microcentrifuge tubes at –20°C for up to 6 mo.

5. ND96: 96 m$M$ NaCl, 2 m$M$ KCl, 1 m$M$ $MgCl_2$, 5 m$M$ HEPES, 1.8 m$M$ $CaCl_2$, pH 7.6. Mix in the following order: 900 mL of distilled sterile water, 50 mL ND96 Ø $Ca^{2+}$ high salt stock, 1.8 mL of 1 $M$ $CaCl_2$.

6. NDE: 96 m$M$ NaCl, 2 m$M$ KCl, 1 m$M$ MgCl$_2$, 5 m$M$ HEPES, 1.8 m$M$ CaCl$_2$, 2.5 m$M$ pyruvate, 50 μg/mL gentamycin sulfate, pH 7.6. Mix in the following order: 900 mL of distilled sterile water, 50 mL ND96 Ø Ca$^{2+}$ high salt stock, 1.8 mL of 1 $M$ CaCl$_2$, 200 μL of gentamycin stock and 2.5 mL of pyruvate stock.

7. ND96 Ø Ca$^{2+}$: 96 m$M$ NaCl, 2 m$M$ KCl, 1 m$M$ MgCl$_2$, 5 m$M$ HEPES, pH 7.6. Add 50 mL ND96 Ø Ca$^{2+}$ high salt stock to 900 mL of distilled sterile water.

8. Check pH of the solutions described in **items 5, 6**, and **7**, and if necessary adjust at 7.6, and then add enough distilled sterile water to 1 L. Store at 4°C. Warm up at 18°C before use.

9. MBS high salt stock: 128 g NaCl, 2 g KCl, 5 g NaHCO$_3$, 89 g HEPES. Dissolve in 800 mL of distilled sterile water and adjust the pH to 7.6 with 1 $M$ NaOH. Make up to 1 L and store as 40-mL aliquots at –20°C.

10. Divalent cation stock: 1.9 g Ca(NO$_3$)$_2$.4H$_2$O, 2.25 g CaCl$_2$.6H$_2$O, 5 g MgSO$_4$·7H$_2$O. Dissolve in 1 L of distilled sterile water and store as 40-mL aliquots at –20°C.

11. Antibiotic stocks: 10 mg/mL sodium penicillin, 10 mg/mL streptomycin sulfate, 100 mg/mL gentamycin sulfate, all dissolved in distilled sterile water and stored at –20°C.

12. Modified Barths' saline (MBS): 88 m$M$ NaCl, 1 m$M$ KCl, 2.4 m$M$ NaHCO$_3$, 15 m$M$ HEPES-NaOH pH 7.6, 0.30 m$M$ Ca(NO$_3$)$_2$, 0.41 m$M$ CaCl$_2$, 0.82 m$M$ MgSO$_4$, 10 μg/mL sodium penicillin, 10 μg/mL streptomycin sulfate, 100 μg/mL gentamycin sulfate. Mix in the following order: 917 mL of distilled sterile water, 40 mL high salt stock, 40 mL divalent cation stock, 1 mL each antibiotic stock. Store at 4°C and warm up to 18°C before use.

13. Collagenase in ND96 Ø Ca$^{2+}$: add 1 g of collagenase type Ia (Sigma) to 1 L of sterile ND96 Ø Ca$^{2+}$. Under a laminar flow hood, make 5-mL aliquots in 16-mm diameter, 100-mm long sterile round-based tubes with screw cap. Store at –80°C for up to 1 yr, or at –20°C for up to 3 mo (*see* **Note 4**).

14. Phage polymerase 20 U/μL. Transcription buffer (without DTT) and DTT stocks are normally supplied together with the polymerase.

15. 10 m$M$ m$^7$G(5')ppp(5')G (Cap analog, Promega).

16. NTP mix (5 m$M$ ATP, 0.5 m$M$ GTP, 5 m$M$ CTP, 5 m$M$ UTP).

17. 25 m$M$ GTP.

18. RNase inhibitor (RNasin, Promega).

19. 8 $M$ LiCl. DNase- and RNase-free (Sigma, Molecular Biology Grade).

20. PRO-MIX L-[$^{35}$S] in vitro Cell Labeling Mix (GE Healthcare). Contains approx 70% L-[$^{35}$S] methionine and 30% L-[$^{35}$S] cysteine. Specific activity L-[$^{35}$S] methionine >1000 Ci/mmol.
21. Homogenization buffer: 20 m$M$ Tris-HCl, pH 7.6, 0.1 $M$ NaCl, 1% Triton X-100. Store at –20°C. Just before use, add protease inhibitor cocktail (Complete, Roche Diagnostics).
22. NET buffer: 50 m$M$ Tris-HCl, pH 7.5, 150 m$M$ NaCl, 1 m$M$ EDTA, 0.1% Igepal-CA630, 0.02% sodium azide. Store at 4°C.
23. NET-gel buffer: NET-buffer supplemented with 0.25% gelatin. Store at 4°C. A 2.5% gelatin stock can be prepared by autoclaving 25 g gelatin in 1 L of distilled water.
24. 10% Protein A Sepharose CL-4B (GE Healthcare) slurry equilibrated with NET buffer. Store at 4°C.

## 3. Methods

### 3.1. Removal of Oocytes

Depending on the number of oocytes required the frog can be subjected to a partial removal of the ovary under anesthesia or can be subjected to euthanasia for removal of the complete ovary. Experiments requiring up to 1000 single oocytes should not normally require the removal of the whole ovary.

Although it is not necessary to use aseptic surgery technique, this is recommended by some authors and has been reported to improve oocyte quality *(4)*. To reduce contamination, we expose the room to ultraviolet light for 2 h to overnight before surgery and before injection (we particularly recommend to follow this practice when the room where oocytes are prepared and injected is close to a molecular biology or microbiology laboratory).

The procedure should be in accordance with local legislation and conform to contemporary standards of care and use of experimental animals. Supervision from an experienced person would be helpful initially.

### 3.1.1. Partial Ovary Removal

A mature, large *Xenopus* female is anesthetized by immersion in a 0.1% solution of 3-aminobenzoic acid ethyl ester (methane sulfonate salt, Sigma; also called Tricaine or MS-222). After about 5 min start checking the frog by gently squeezing the toes with the fingers, and repeat until no response is elicited. Less that 20 min should normally suffice. The frog should remain anaesthetized for about 20 min. The anesthetized frog is then placed on its back on ice covered with wet paper, and covered with wet paper leaving only the abdominal area off. Keep the skin of the frog well moistened with water throughout the procedure.

1. After elevating the skin at a position away from the midline (which contains a major blood vessel) in the lower abdominal quadrant, make a 10-mm incision using a pair of small scissors (**Fig. 1**). This exposes the body wall, which is then cut (0.75 cm) using the same technique. The ovary is now exposed.
2. Tease out one or a few lobes of the ovary, using forceps, and excise them using the scissors.
3. Immediately transfers the lobes to MBS (or ND96 Ø $Ca^{2+}$ if oocytes will be enzymatically defolliculated).
4. Repeat **steps 2** and **3** until enough oocytes are obtained. Each ovary consists of about 24 lobes.
5. Suture the incision in the body wall separately from that in the skin, using two stitches of absorbable synthetic thread (Dar-Vin 4/0) for the former and three of silk thread (4/0) for the latter.
6. Leave the frog for the first few hours after surgery in a small tank with 2–5 cm of water until a complete awakening.

### 3.1.2. Complete Removal of the Ovary

Euthanasia can be performed by immersing the animal in an overdose of anesthetic (>0.3%). During the dissection it is important to avoid damage to the major abdominal blood vessels. We therefore suggest to follow the same procedure as previously described for the partial removal of the ovary.

### 3.2. Oocyte Isolation (see Note 5)

1. Soon after their removal from the frog, thoroughly wash the ovary lobes in MBS and tease them apart, using watchmaker forceps, into clumps of not more than 50 large oocytes.
2. Store the clumps in MBS at a temperature ranging from 14 to 20°C. Lower temperatures prolong the useful life of the oocytes.

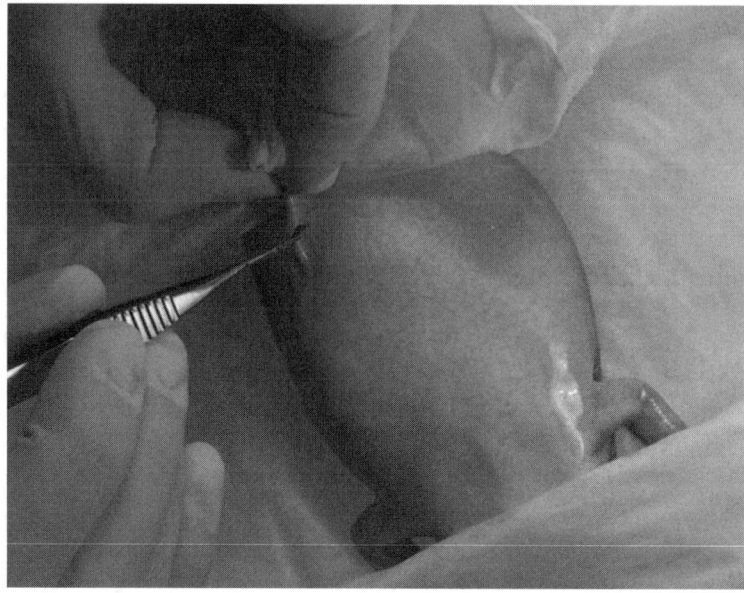

Fig 1. Frog surgery. After lifting the skin, a small incision is made in the lower abdominal quadrant.

3. Transfer an oocyte clump to a Petri dish containing MBS and separate single oocytes manually using forceps or a stainless steel wire loop.
4. Select good, undamaged oocytes under a stereomicroscope and transfer them to fresh medium.

### 3.3. Preparation of Defolliculated Oocytes (see Note 6)

1. Soon after their removal from the frog, thoroughly wash the ovary lobes with ND96 Ø $Ca^{2+}$ in a 90-mm diameter Petri dish.
2. Using watch-maker forceps or small scissors, divide the ovary into clumps of not more than 20 large oocytes and wash them in ND96 Ø $Ca^{2+}$ to eliminate any material released by damaged oocytes (*see* **Note 7**).
3. Transfer the oocyte clumps (for a total of about 200 oocytes) in a tube containing 5 mL of collagenase solution.
4. Gently swirl in the dark at 18–22°C.
5. After the first 30 min, and then every 10 min, carefully check the tubes (*see* **Note 8**). When the oocytes appear well separated (**Fig. 2A**) wash them three to five times with ND96 Ø $Ca^{2+}$, then three to five times with ND96 and twice with NDE before transferring them to a 90-mm Petri dish

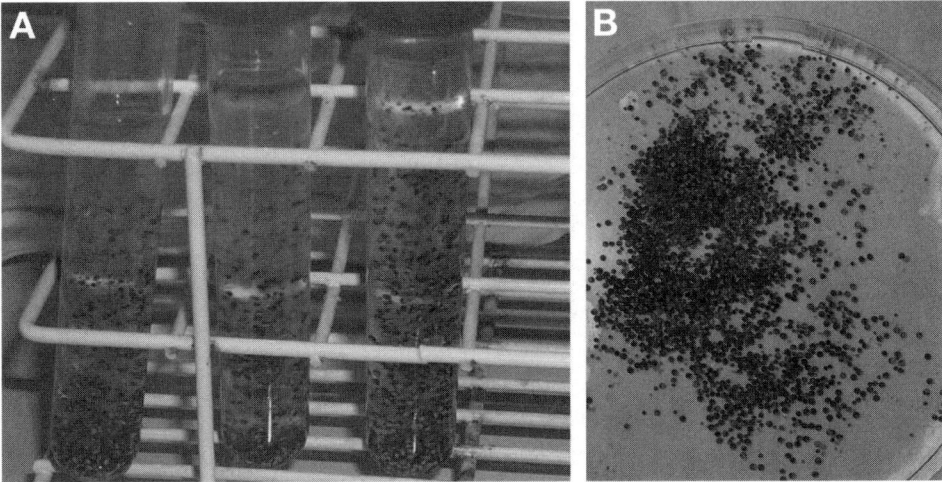

Fig. 2. (**A**) Collagenase treatment. (**B**) After collagenase treatment the oocytes should look well isolated.

in NDE solution (*see* **Note 9**). The oocytes now will look well separated in a clean solution free of particulate matter (**Fig. 2B**).

6. Select good stage V-VI oocytes and transfer them into fresh medium.
7. Store the isolated oocytes in 60-mm diameter Petri dishes at 14–20°C in NDE.

### 3.4. RNA Preparation

Either natural or synthetic RNAs can be injected into *Xenopus* oocytes. If natural RNA is used (and if the mRNA of interest is polyadenylated) it is advisable to eliminate most of the ribosomal RNA using one of the standard poly(A)+ mRNA purification techniques. Although ribosomal RNA does not interfere with mRNA translation, the enrichment of poly(A)+ mRNA can be important when the mRNA of interest does not represent a major component of the mRNA population. We routinely store RNA at –80°C in distilled water at twice the concentration that will be used for injections.The following protocol can be used for the preparation of large amounts of cRNA (*see* **Note 10**).

1. Linearize 5–10 μg of plasmid for 2 h or overnight with 20 U of an appropriate restriction enzyme, in a final volume of 50 μL.

2. Bring the volume to 100 µL with double-distilled water.
3. Purify the digested plasmid using the Wizard® SV Gel or PCR Clean-Up System (Promega). Other PCR purification kits should be suitable.
4. Elute the plasmid in 35 µL RNase free double-distilled water.
5. Take 32 µL of the eluted DNA solution and add: 18 µL 5X Trascription Buffer, 8 µL of DTT 100 mM, 2.5 µL of RNasin 30 U/µL, 13 µL NTPs mix, 6.5 µL 10 mM m$^7$G(5')ppp(5')G (Cap Analog), 10 µL RNA polymerase 20 U/µL (final volume 90 µL).
6. Incubate at 37°C. After 10, 20, and 40 min from the start of the transcription reaction, add 1 µL of 25 mM GTP.
7. After 1 h from the start of the transcription reaction add 4 µL 5X transcription buffer, 1 µL of 100 mM DTT, 1 µL RNasin 30 U/µL, 5 µl NTPs mix, 1 µL RNA polymerase 20 U/µL, 1 µL 25 mM GTP, 4 µL of RNase-free double distilled water (final volume 110 µL).
8. Incubate for 2–3 h at 37°C.
9. Add 90 µL of RNase-free double distilled water to obtain a final volume of 200 µL.
10. Add 1 vol of phenol:chloroform:isoamyl alcohol, 25:24:1, pH 6.6 (Ambion) and vortex for 1'.
11. Centrifuge for 5 min at 10,000$g$.
12. Transfer the top phase into a fresh microcentrifuge tube.
13. Add 1 vol of sevag (chloroform:isoamyl alcohol 24:1).
14. Vortex for 1 min.
15. Centrifuge for 5 min at 10,000$g$.
16. Transfer the top phase in a fresh microcentrifuge tube.
17. Add 0.75 vol of RNase-free 8 M LiCl. Mix by pippetting.
18. Invert the tube three to four times.
19. Place the tube at −20°C for a minimum of 2 h.
20. Centrifuge for 30 min at 10,000$g$ at 4°C.
21. Carefully remove the supernatant avoiding disturbing the pellet.
22. Wash the pellet with 200 µL of 70% ethanol.
23. Dry the pellet for 5 min.
24. Add 16 µL of RNase-free double-distilled water.
25. Estimate cRNA concentration using a spectrophotometer (1 $A_{260nm}$ unit = 40 µg/mL). More than 30 µg cRNA are routinely synthesized using this protocol.
26. Dilute the cRNA to the desired concentration and split into aliquots. Store at −80°C.

## 3.5. Micropipet Preparation

Exact conditions for micropipet preparation must be determined experimentally following the instructions provided by the manufacturer of the micropipet puller. It will be then necessary to break the tip of the micropipet to obtain an inner diameter (ID) of 5–10 µm.

Procedure:

1. Place the micropipet on a carefully cleaned stereomicroscope stage, and break the tip using the side wall of a 35-mm Petri dish lid, or a pair of watch-maker forceps.
2. Completely fill the micropipet with paraffin oil using a silicon tubing fitted on a 2-mL syringe. Carefully check that no bubbles are present in the tube before connecting it to the micropipet. To avoid bubble formation, keep pressure on the plunger until you have disconnected the micropipet from the silicon tubing (**Fig. 3A**).
3. Fit the micropipet on the microinjector plunger, following the instructions of the microinjector manufacturer (*see* **Note 11**). Discharge paraffin until the piston is at 60% into the micropipet. At all stages of the injection procedure, avoid moving the piston to the end of its stroke. This will allow you to reload the same capillary several times. A microinjection setup is shown in **Fig. 3B**.
4. Seal with Parafilm the base of a 35-mm Petri dish and place it on the microscope stage. Place 2–5 µL of RNA solution onto the Parafilm (*see* **Note 12**). The amount of RNA sample to be loaded into the micropipet will depend on the number of oocytes to be injected and on the injection volume (usually 20–50 nL/oocyte, but up to 70 nL/oocyte can be injected).
5. Insert the tip of the micropipet into the droplet, and start loading the RNA solution (**Fig. 3C**). During the loading process carefully and constantly check the micropipet both with the naked eye and under the microscope. Clogging of the micropipet and bubble formation are constant hazards.

## 3.6. Microinjection Procedure

1. Cement a sheet of polyethylene mesh containing 1 mm$^2$ holes to the bottom of 60-mm Petri dish using a glue stick and a hot-melt glue gun. After careful cleaning, the Petri dish can be reused several times.
2. Fill the Petri with oocyte culture medium, and place the oocytes in a line on the grid, possibly with the vegetative pole up (*see* **Note 13**). Initially prepare for injection not more than 10–20 oocytes.

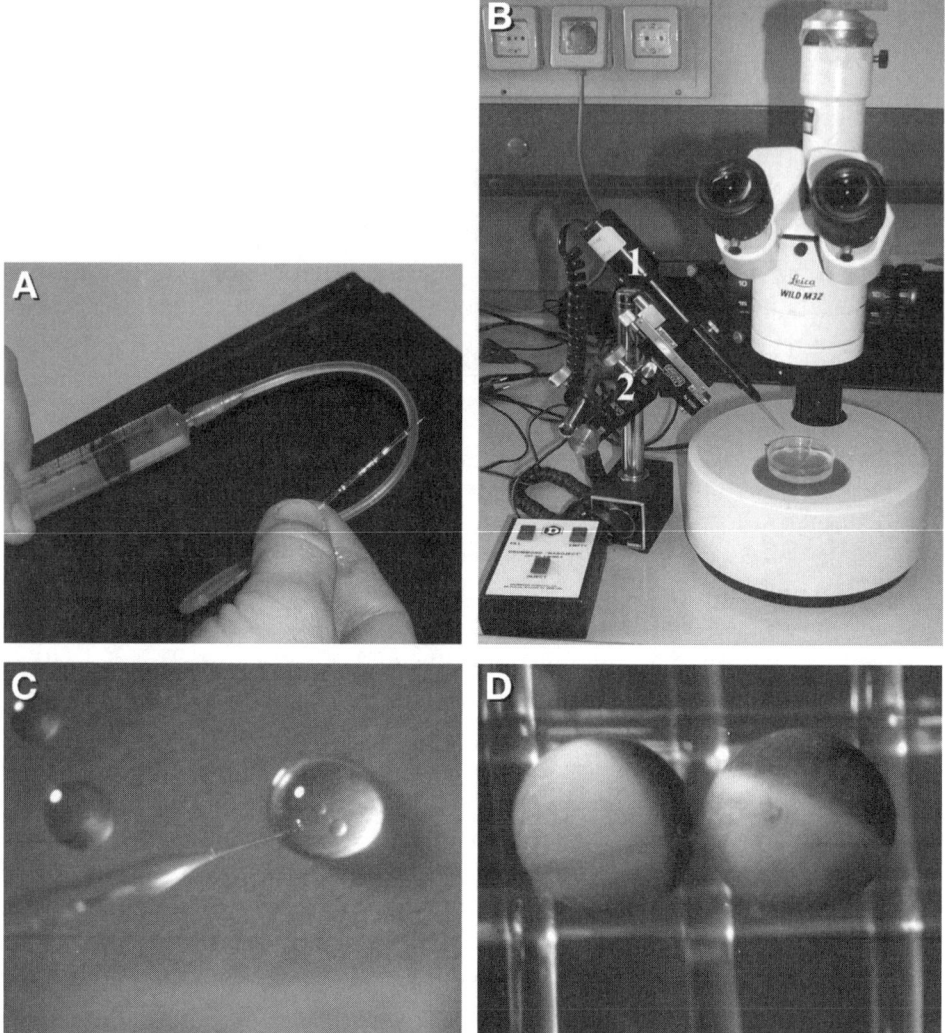

Fig. 3. **(A)** Filling the micropipet with paraffin oil. **(B)** A microinjection setup. The micronjector (1) is fitted on the micromanipulator (2) and placed on the side of the microscope. **(C)** Micropipet loading. **(D)** Oocyte injeciton.

3. Working on the micromanipulator, place the micropipet tip as close as possible to the first oocyte, until you can see a small depression on the oocyte surface. Give a small tap on the back of the injector to insert the micropipet tip into the oocyte (with practice, this step can be omitted and the tip of the micropipet inserted into the oocyte using the controls on the

micromanipulator) and push the injector button (**Fig. 3D**). The oocyte will swell. Wait 1 s and then retract the micropipet tip (*see* **Note 14**).

4. Move the micropipet to the next oocyte and repeat the above procedure. With practice, you will find convenient to move only the Petri dish to place the next oocyte in the right position, and will use the micromanipulator only to move the micropipet up and down.

5. Transfer injected oocytes to a 35-mm Petri dish containing oocyte culture medium, and incubate them at 18°C.

Check the oocytes a few hours after injection, and then at least daily, removing those that do not look healthy. Replace the incubation medium every 24 h.

## 3.7. Detection of the Expressed Protein

*Xenopus* oocytes synthesize about 20 ng of endogenous protein per hour and the expression of the heterologous protein, which can represent a substantial fraction of total protein synthesis, can be normally monitored using a single or few oocytes. This can be done using different procedures such as a bioassay or, if a suitable antibody/antiserum is available, an immunochemical technique (*see* **Note 15**). In this regard it should be noted that not all the oocyte synthesized protein may be functional, and that therefore functional and immunological assays cannot be considered equivalent.

### 3.7.1. Labeling With Amino Acids (see **Note 16**)

To perform a typical pulse-chase experiment, proceed as follows:

1. Select healthy oocytes injected the previous day and transfer them to sterile microfuge tubes (up to 20 oocytes/tube) or to the wells of a 96-well microtiter plate (up to 5 oocytes/well). A total of 10–20 oocytes are normally used for each time-point.

2. Completely remove the medium.

3. Add MBS (5–10 μL/oocyte) containing 0.5–1 μCi/μL radiolabeled methionine and cysteine (e.g., PRO-MIX L-[$^{35}$S] in vitro Cell Labeling Mix, GE Healthcare). The oocytes should be completely submerged (*see* **Note 17**).

4. Incubate for the desired time (normally 1–6 h) at 18–20°C.

5. At the end of the labeling period remove and save the culture medium and wash twice the oocytes with excess fresh MBS. Completely remove the medium and add MBS containing 1 m$M$ unlabeled methionine and 1 m$M$

unlabeled cysteine (5–10 µL per oocyte). In the case of long incubations save and replace daily the incubation medium.

6. At the end of the chase period, save the culture medium and wash the oocytes as previously mentioned. The oocytes can be immediately homogenized or frozen in dry ice.

### 3.7.2. Oocyte Homogenization and Immunoprecipitation of Radiolabeled Proteins

The following protocol gave good results with different soluble and membrane proteins expressed in *Xenopus* oocytes. The reader can refer to **ref. 5** for a detailed discussion of immunoprecipitation procedures.

Perform all steps at 4°C.

1. Transfer fresh or frozen oocytes to a cold loose-fitting glass homogenizer. Add 40 µL of homogenization buffer per oocyte and homogenize.
2. Transfer the sample to a 1.5-mL microcentrifuge tube and spin for 5 min at 10,000$g$. This will pellet most (but not all) of the storage proteins of the oocyte. A thin layer of lipid will also be evident at the surface.
3. Remove most of the supernatant, trying to avoid the lipid pellicle, and transfer it to a fresh tube. Five microliters of this extract can be analyzed by sodium dodecyl sulfate-polyacrylamide gels (SDS-PAGE).
4. Remove an aliquot of the clarified homogenate or culture medium (typically one oocyte equivalent) and transfer it to a fresh 1.5-mL microcentrifuge tube (*see* **Note 18**).
5. Add NET-gel buffer to a final volume of 1 mL.
6. Add appropriate antibody (e.g., 1–5 µL of a polyclonal antiserum).
7. Mix and incubate for 2 h on ice.
8. Add 100 µL of a 10% Protein A-Sepharose CL-4B bead slurry (*see* **Note 19**).
9. Mix thoroughly by inversion and rotate tubes end over end for more than 1 h at 4°C.
10. Microcentrifuge 1 min at 6000$g$.
11. Remove and discard the supernatant, leaving about 10 µL on top of beads.
12. Add 1 mL of NET-gel and mix by inversion.
13. Microcentrifuge 1 min at 6000$g$.
14. Remove supernatant leaving about 10 µL on top of beads.
15. Wash beads two more times with NET-gel, completely removing the supernatant after the last wash.
16. Store beads at –80°C before analysis by SDS-PAGE.

## 4. Notes

1. Sexually mature females and extra large females usually give oocytes of comparable quality. The laboratory conditioned and the laboratory reared animals are used to pelleted food. Some suppliers will also ship extracted ovaries.

2. Before obtaining and utilizing *X. laevis* in the laboratory, the relevant information about any required permit, license, and authorization concerning the use of these animals should be obtained.

   Adult *X. laevis* can be maintained on a 12 h light/dark cycle in dark plastic tanks containing 10–15 cm of water, allowing at least 2.5–3 L/animal. The walls of the tank should be high enough (20 cm above water level) to discourage the frogs from attempting to escape. The top of the tank should be firmly closed with a lid made of plastic or metallic mesh that will ensure good ventilation. The presence of some hiding places, such as some PVC half pipe tube of 20 cm of diameter, would be appreciated by the frogs.

   The quality of the water is crucial to the health of the frogs and hence of the oocytes. Different values for parameters like pH, temperature, conductivity, dissolved oxygen, general hardness, carbonate hardness are reported in the literature *(4,6–8)*: pH from 6.5 to 8.5; temperature from 17 to 23°C, conductivity 500–2000 µS; dissolved oxygen 3–4 ppm, general hardness from 40 to 500 ppm, carbonate hardness from 20 to 400. Untreated tap water may be suitable so long as it is hard (i.e., high in $Ca^{2+}$), unchlorinated and very low in organics and heavy metal ions. Chlorine can be lowered to an acceptable level by standing the water for at least 24 h in an open tank but if it contains chloramine, the water should be filtered through a carbon filter, or treated with a suitable powder or pellet. Information about tap water data can be obtained from the local water district office.

   If a static system is used water should be changed after each meal, or more frequently if obviously needed. Static systems are suitable for small colonies. If a large number of animals must be housed, it is convenient to use a continuous recirculating system. The ultimate, though expensive systems is a complete recirculating shelf system with mechanical, biological, and chemical filtration, plus ultraviolet sterilization. The rack utilizes a control system to monitor and automatically adjust dissolved oxygen, pH, salinity, temperature, and conductivity (Aquatic Habitats, www.aquatic habitats.com; Tecniplast, www.tecniplast.it).

A cheaper alternative, the one we use, is a continuously recirculating system with a chemical and biological filter. You will need to carefully monitor the water in this system as you are continuously reusing it, and to make sure that the system is maintained perfectly efficient because any problem in one tank would inevitably spread to other tanks. In summary, static systems are inexpensive, work well but have the disadvantage to need constant manual cleaning. Recirculating systems using biological and mechanical filters need constant maintenance and a perfect control of the quality of the water.

The frog tanks should be placed in a quiet room in which the temperature is constantly maintained between 18 and 22°C. Noise and temperatures >23°C generate stress in the animals, affecting oocyte quality.

The frogs recovering from surgery should be kept in a separate and well-monitored tank for 1 wk.

*X. laevis* is carnivorous, and although different diets can be appropriate, they should be well balanced and complete. We prefer to feed the frogs twice a week using special dry food (Frog Brittle, Nasco) or trout or salmon chow (AquaMax, Purina). Usually 10 pellets per adult frog are sufficient. Animals should be observed during feeding to monitor any abnormal behavior.

3. Infectious diseases of frogs. Disease is not usually a serious problem and might be an indicator of environmental stress. The most common problem is a bacterial infection called red-leg. The skin becomes rough and red through haemolysis. Because of the breakdown in osmotic regulation the frogs should be placed in 0.5% NaCl (this is useful whenever the surface of the animal is damaged). Treatment by immersion in oxytetracyclin (50 μg/mL for 2 wk) has been reported to be efficacious. Affected animals are always isolated from the main tanks. If only a few animals are affected it is best to eliminate them.

Another disease, called "flaky skin disease" is caused by the nematode *Capillaria xenopodis* or *Pseudocapillariodes xenopi*. This infection causes epidermal desquamation and can favor the development of the "red leg" disease. It can be effectively treated by leaving the frogs overnight in 100 μg/mL thiabendazole (Sigma). We find that this treatment often results in the cure of animals with "red leg" disease. The successful use of levamisole, which is better tolerated by the frogs, has also been reported *(9)*.

4. Before ordering the collagenase, we recommend to check the certificate of analysis and in particular to control collagen digestion activity that should not exceed 250/300 U/mg and FALGPA hydrolysis activity that

must be lower than 2 U/mg solid. If the activity is too high the action of the collagenase on the oocyte will be too fast and the oocytes will be damaged by the enzymatic treatment without loosing the follicular layer. Check each new batch of collagenase by testing it on oocytes. Then purchase or ask to reserve for you a large amount of enzyme of the same batch.

Although $Ca^{2+}/Mg^{2+}$-free MBS can also be used as a medium for collagenase treatment, ND96 $\emptyset$ $Ca^{2+}$ is the medium we routinely use in our laboratory.

5. It is generally better to isolate oocytes the day before injection. This allows any damage to the oocyte owing to manipulation to become evident, and to remove such damaged oocytes. Isolated oocytes are handled using a shortened Pasteur pipet. The mouth should be 2–3 mm wide. Store the isolated oocytes in 3.5-cm diameter Petri dishes at 14–20°C. Although culturing of oocytes does not require stringent sterile technique, it is advisable to minimize contamination of the culture medium by using sterile glassware and plasticware.

In *Xenopus*, oogenesis is asynchronous and the lobes will contain oocytes in all stage of development. According to Dumont *(10)* the development of *Xenopus* oocytes can be divided into six stages (I to VI) **(Fig. 4)**. At stage IV, an animal (dark) and a vegetal (yellow) hemisphere are well differentiated. In stage V oocytes (1000–1200 mm diameter) the color of the animal hemisphere turns brown or beige. Stage VI oocytes (>1200 mm diameter) are sometimes distinguished by the presence of a relatively unpigmented equatorial band (0.2 mm wide). For a detailed description of oogenesis in *X. laevis see* Smith et al. *(11)*. Because only stage V and stage VI oocytes are normally utilized for injection of mRNA, it is important that these stages are well represented within the oocyte population. It is not unusual to see a small portion of large oocytes which are undergoing atresia (death and readsorption).

Oocytes can be easily damaged during their separation from the ovary or (at much lower frequency) during their microinjection. Sick oocytes can be easily recognized among healthy ones by the presence of uneven patches of pigment on their surface, particularly evident when located in the animal half **(Fig. 5A,B)**. Sometimes the oocytes look healthy but are extremely turgid as a result of a loss in osmotic regulation; such oocytes bounce off probing forceps. It is important to recognize and remove all these oocytes before they release large amounts of cytosolic components into the incubation medium.

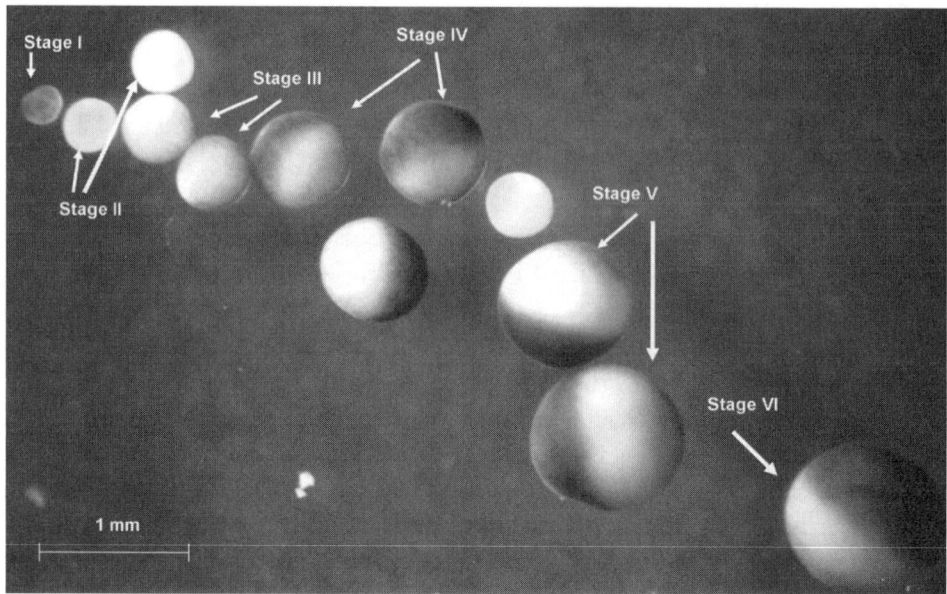

Fig. 4. Oocytes at different stages of development.

6. Manually isolated oocytes are often surrounded by their thecal layer, follicle cells, and vitelline envelope (**Fig. 6**). The thecal layer is also lost
from enzymatically defolliculated oocytes. The absence of the theca
(which can also be manually removed) makes the oocytes less resilient
but at the same time easier to penetrate with a micropipet and many work-
ers will routinely remove it for this reason alone.

It is necessary to remove the follicular layer when the oocytes are used for
electrophysiological experiments *(12)*, first because only defolliculated
oocytes can be injected and impaled without damage, second because of
the receptors, channels, and transporters endogenously present in the follicular cells that can interfere with the registration of the current induced
by the heterologous protein expressed by the oocytes.

The easy way to test whether the oocytes are defolliculated is to observe
them under lateral illumination. The defolliculated oocytes have a distinct
glare owing to the reflection of the light by the vitelline membrane (**Fig. 7**).
If the collagenase treatment was not sufficient, the residual follicular layer
can be removed by passing the oocytes up and down in a fire-polished
Pasteur pipet slightly smaller than the oocyte. Finally, it should be noted
that collagenase treatment has been reported to cause a transient depression of endogenous protein synthesis *(11)*.

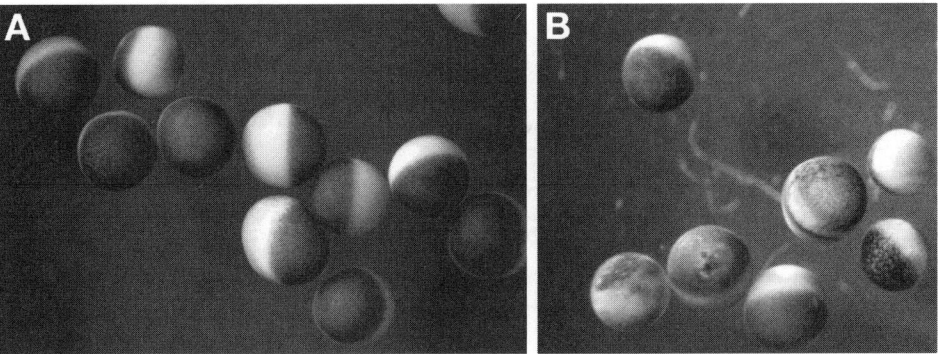

Fig. 5. **(A)** Good oocytes. **(B)** Sick oocytes.

7. Healthy oocytes can be damaged by the proteolytic enzymes released by damaged oocytes. These enzymes may digest the membrane during collagenase treatment, and can be a source of potassium that can depolarize the oocytes.

8. Especially when a new collagenase batch is used, it is imperative to frequently check the condition of the oocytes. If the collagenase solution becomes too dirty and opaque, replace it with a new aliquot.

9. These steps can be conveniently performed with a Pasteur pipet connected to a vacuum flask.

10. In vitro transcripts can be conveniently used to obtain the expression of cloned genes in *Xenopus* oocytes. It has been shown that the presence of a monomethyl cap ($m^7G(5')ppp(5')G$) and a poly(A) tail greatly enhance both stability and translational efficiency of the microinjected transcript *(13–15)*. These features should therefore be incorporated into the synthetic transcript.

    The 5'-cap structure can be most conveniently introduced by supplying the transcription reaction with the appropriate cap analogue ($m^7G(5')ppp(5')G$; Promega). The transcription product can be purified from the reaction components with a phenol-chloroform extraction followed by LiCl precipitation. The use of vectors with SP6 or T7 promoters is recommended, because T3 polymerase has been reported not to produce stable capped transcripts *(16)*.

    A convenient way to add a poly(A) tail is to have it synthesized by the RNA polymerase as part of the transcription product. This can be obtained by cloning the desired coding sequence into a vector like pSP64T *(15)*, pSP6TN *(17)*, or pOO2 *(18)*. These vectors offer the additional advantage

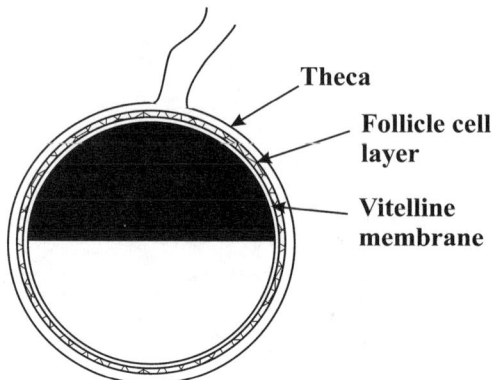

Fig. 6. Schematic drawing showing an oocyte surrounded by the vitelline membrane, the follicle cell layer and the theca.

of placing any coding sequence within the 5'- and the 3'-untranslated regions of an efficiently translated *Xenopus* gene (β-globin). Similarly, the pAMV-PA *(19)* vector provides the transcript with an 5'-untranslated region derived from the alfalfa mosaic virus and a poly(A) tail. These vectors contain the promoter sequence recognized by a phage (SP6 or T7) RNA polymerase. The use of pSP64T as transcription vector has also been reported to give very consistent expression levels in oocytes from the same or from different frogs *(20)*. We find that the protocol described here consistently produce efficiently translated transcripts with good yields, but commercial in vitro transcription kits can also be used (mMESSAGE mMACHINE, Ambion).

11. Before assembling the injector, carefully clean with ethanol the displacement piston, the o-rings, and the plastic sleeve to remove any residual paraffin and ensure a perfect seal.

12. Just before use, centrifuge the RNA solution for 20 s at top speed in a microfuge to sediment any debris that could block the micropipet.

    In practice, the injection of 50 nL/oocyte of a 0.2 mg/mL mRNA solution of an in vitro transcript or of a purified mRNA species and of 50 nL/oocyte of a 2 mg/mL preparation of total poly (A)+ mRNA can be considered routine. Injection of high amounts (50 nL of a 10 mg/mL solution) of RNA does not normally produce any detectable toxic effect. We should point out however that because the translational efficiency of different mRNAs shows considerable variation, it is advisable to try different concentrations of a particular mRNA.

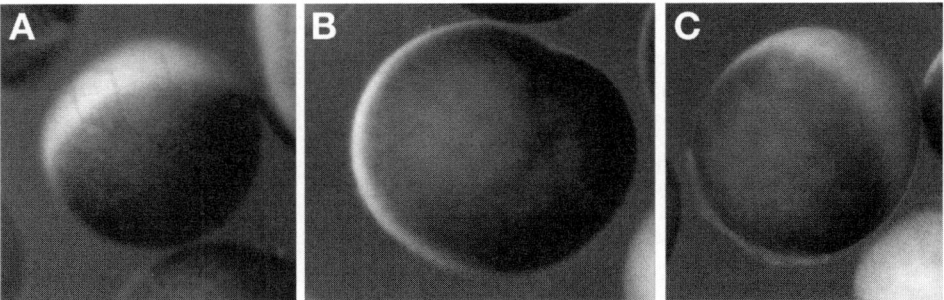

Fig. 7. (**A**) A folliculated oocyte. Capillaries are often evident on the background of the vegetal hemisphere. (**B**) A partially defolliculated oocyte. Note the constriction owing to partial follicle removal. (**C**) A defolliculated oocyte.

13. In most cases, the position where the mRNA is deposited is of little relevance to the outcome of the experiment and the interpretation of results. Although the animal hemisphere contains 60% of the ribosomal RNA, injection at the animal pole does not seem to result in enhanced translation *(21)*. Rather, it is important to avoid depositing the RNA into the nucleus, that resides in the animal hemisphere. We therefore suggest to inject RNA into the vegetal half. In general, injected natural mRNAs diffuse slowly from the site of injection *(21)*. Natural mRNAs deposited at the animal pole can be expected to be still mostly localized in the animal half even 48 h after the injection. Diffusion in the opposite direction is instead more rapid. Certain proteins synthesized from the injected RNA have been shown to diffuse from their site of synthesis *(21,22)*, but others have been found to remain largely localized where they are synthesized *(22,23)*. These observations may be of some relevance for certain specific applications.

14. A constant hazard in microinjection is the blocking of the micropipet. This can occur from particles in the RNA sample or from oocyte yolk particles sucked back into the micropipet. The best way of dealing with a blocked needle is to break off the tip beyond the blockage with heat-treated forceps. So long as the diameter of the new tip is less than 15 μm, injection can be continued. To monitor the reproducibility of the injection procedure the oocytes can be injected with cRNA encoding a secreted alkaline phosphatase, and phosphatase activity in the oocyte medium can be then measured using a standard colorimetric assay *(24)*. Alternatively, a simple way to test whether you have achieved competence is to inject oocytes with a 0.1 mg/mL solution of rabbit globin mRNA (commercially avail-

able) and then label the oocytes for a few hours as described in **Subheading 3.7.1.** After individually homogenizing the oocytes, a volume corresponding to 1/10 of each injected oocyte (and of water-injected ones, as control) can be directly analyzed by SDS-PAGE. The globin band runs slightly anomalously relative to protein markers at about 13 kDa and is easily seen above a background of endogenous proteins.

15. The presence of the abundant yolk proteins may represent a problem if the oocyte homogenate has to be directly analyzed by Western blot. Procedures to reduce the presence of yolk proteins in the oocyte protein extract have been reported *(25,26)*

16. It is advisable to incubate mRNA-injected oocytes for several hours prior to label addition, to allow the identification and elimination of damaged oocytes. The presence of more than 10% of sick oocytes postinjection generally means that the injection has not been performed properly or that oocyte quality is substandard. The material released by degenerating oocytes can constitute a substrate for growth of bacteria and fungi, an undesired effect especially when the secreted medium has to be analyzed.

17. Although we have traditionally performed pulse-chase assays in MBS, other media can be used. Labeled amino acids can be either added to the medium or injected. The specific activity of the supplied amino acid, the size of the endogenous pool, and the number of residues present in the polypeptide chain will influence the efficiency of labeling. The use of methionine/cysteine mixtures combines the advantages of relatively low cost, high specific activity, small endogenous pools, and strong $\beta$ emission. If your protein does not contain methionine or cysteine residues, [$^3$H]leucine (1µCi/µL) is a possible alternative. If, as in the case of [$^{35}$S]methionine/cysteine mixtures and [$^3$H]leucine, the precursor can be purchased at concentrations of several µCi/µL in aqueous solution, it is appropriate to add to the radiolabeled amino acid solution an equal volume of 2X MBS and then dilute the solution with MBS to the desired final concentration. Otherwise, the amino acid solution is dried down in a Savant Speedy-Vac and resuspended in MBS. For injection, the radioactive amino acid is dried as previously listed in a siliconized 0.5-mL microfuge tube and then resuspended in water at high concentration (e.g., 50 µCi/µL). Fifty nanoliters of this solution can be injected into each oocyte using the same technique described for RNA injection. This injection is usually subsequent to RNA injection although mRNA and label can be co-injected, if desired.

If the labeling period has to be followed by a chase period, the oocytes are transferred to MBS containing 1 m$M$ of the appropriate unlabeled amino acid. If the medium containing the radioactive precursor is not removed the oocytes will utilize most of the supplied amino acid within the first few hours. We find that the incorporation of [$^{35}$S]methionine (>1000 Ci/mmole) or [$^3$H]leucine (about 150 Ci/mmol) is normally restricted to the first 8 h of incubation.

If the incubation medium has to be analyzed for the presence of secreted proteins, it is generally advisable to first saturate protein binding sites present on the microtiter wells or microfuge tubes treating the wells or tubes for 20 min with a 0.5% solution of bovine serum albumin in MBS, before rinsing with MBS.

18. For analysis of secreted proteins, clarify the incubation medium for 5 min at 10,000$g$.
19. Use Protein-G beads if the antibody belongs to a species or to a subclass that does not bind Protein A.

# References

1. Fabbrini, M. S., Carpani, D., Soria, M. R., and Ceriotti, A. (2000) Cytosolic immunization allows the expression of preATF-saporin chimeric toxin in eukaryotic cells. *FASEB J.* **14,** 391–398.
2. Ceriotti, A. and Colman, A. (1990) Trimer formation determines the rate of influenza virus haemagglutinin transport in the early stages of secretion in *Xenopus* oocytes. *J. Cell Biol.* **111,** 409–420.
3. Blumenstein, Y., Ivanina, T., Shistik, E., Bossi, E., Peres, A., and Dascal, N. (1999) Regulation of cardiac L-type Ca$^{2+}$ channel by coexpression of G$_{as}$ in *Xenopus* oocytes. *FEBS Lett.* **444,** 78–84.
4. Schultz, T. W. and Dawson, D. A. (2003) Housing and husbandry of *Xenopus* for oocyte production. *Lab. Anim.* **32,** 34–39.
5. Bonifacino, J. S. (2000) Protein labelling and immunoprecipitation, in *Current Protocols in Cell Biology, Cell Biology and Cytology,* (Bonifacino, J. S., ed.), Wiley, Hoboken, NJ.
6. Green, S. L. (2002) Factors affecting oogenesis in the South African clawed frog (*Xenopus laevis*). *Comp. Med.* **52,** 307–312.
7. Godfrey, E. W. and Sanders, G. E. (2004) Effect of water hardness on oocyte quality and embryo development in the African clawed frog (*Xenopus laevis*). *Comp. Med.* **54,** 170–175.

8. Hilken, G., Dimigen, J., and Iglauer, F. (1995) Growth of *Xenopus laevis* under different laboratory rearing conditions. *Lab. Anim.* **29,** 152–162.

9. Iglauer, F., Willmann, F., Hilken, G., Huisinga, E., and Dimigen, J. (1997) Anthelmintic treatment to eradicate cutaneus capillariasis in a colony of South African clawed frogs. *Lab. Anim. Sci.* **47,** 477–482.

10. Dumont, J. N. (1972) Oogenesis in *Xenopus Laevis (Daudin)*. Stages of oocyte development in laboratory maintained animals. *J. Morphol.* **136,** 153–179.

11. Smith, L. D., Weilong, X., and Varnold, R. L. (1991) Oogenesis and oocyte isolation, in *Methods in Cell Biology, Vol. 3, Xenopus laevis: Practical Use in Cell and Molecular Biology* (Kay, B. K. and Peng, H. B., eds.), Academic Press, San Diego, CA, pp. 45–60.

12. Dascal, N. and Lotan, I. (1992) Expression of exogenus ion channel and neurotrasmitter receptor in RNA-injected *Xenopus* oocytes, in *Methods in Molecular Biology, Vol. 13, Protocols in Molecular Neurobiology,* (Bausch, S. and Prevest, eds.), Humana Press, Totowa, NJ, pp. 205–225.

13. Drummond, D. R., Armstrong, J., and Colman, A. (1985) The effect of capping and polyadenylation on the stability, movement and translation of synthetic messenger RNAs in *Xenopus* oocytes. *Nucl. Acids Res.* **13,** 7375–7394.

14. Furuichi, Y., LaFiandra, A., and Shatkin, A. J. (1977) 5'-Terminal structure and mRNA stability. *Nature* **266,** 235–239.

15. Krieg, P. A. and Melton, D. A. (1984) Functional messenger RNAs are produced by SP6 *in vitro* transcription of cloned cDNAs. *Nucl. Acids Res.* **12,** 7057–7070.

16. Leaf, D. S., Roberts, S. J., Gerhart, J. C., and Moore, H. (1990) The secretory pathway is blocked between the trans-Golgi and the plasma membrane during meiotic maturation in *Xenopus* oocytes. *Dev. Biol.* **141,** 1–12.

17. Becq, F., Hamon, Y., Bajetto, A., Gola, M., Verrier, B., and Chimini, G. (1997) ABC1, an ATP binding cassette transporter required for phagocytosis of apoptotic cells, generates a regulated anion flux after expression in *Xenopus laevis* oocytes. *J. Biol. Chem.* **272,** 2695–2699.

18. Ludewig, U., von Wiren, N., and Frommer, W. B. (2002) Uniport of $NH_4^+$ by the root hair plasma membrane ammonium transporter LeAMT1;1. *J. Biol. Chem.* **277,** 13,548–13,555.

19. Nowak, M. W., Gallivan, J. P., Silverman, S. K., Labarca, C., Daugherty, D. A., and Lester, H. A. (1998) In vivo incorporation of unnatural amino acids into ion channels in *Xenopus* oocytes expression system. *Methods Enzymol.* **293,** 504–529.

20. Buller, A. L. and White, M. M. (1988) Control of Torpedo acethylcoline receptor biosynthesis in *Xenopus* oocytes. *Proc. Natl. Aca. Sci. USA* **85,** 8717–8721.

21. Drummond, D. R., McCrae, M. A., and Colman, A. (1985) Stability and movement of mRNAs and their encoded proteins in *Xenopus* oocytes. *J. Cell Biol.* **100,** 1148–1156.

22. Ceriotti, A. and Colman, A. (1988) Binding to membrane proteins within the endoplasmic reticulum cannot explain the retention of the glucose-regulated protein GRP78 in *Xenopus* oocytes. *EMBO J.* **7,** 633–638.

23. Altschuler, Y. and Galili, G. (1994) Role of conserved cysteines of a wheat gliadin in its transport and assembly into protein bodies in *Xenopus* oocytes. *J. Biol. Chem.* **269,** 6677–6682.

24. Tate, S. S., Urade, R., Micanovic, R., Gerber, L., and Udenfriend, S. (1990) Secreted alkaline phosphatase: an internal standard for expression of injected mRNAs in *Xenopus* oocytes. *FASEB J.* **4,** 227–231.

25. Evans, J. P. and Kay, B. K. (1991) Biochemical fractionation of oocytes, in *Methods in Cell Biology, Vol. 36, Xenopus laevis: Pratical Uses in Cell and Molecular Biology,* (Kay, B. K. and Peng, H. B., eds.), Academic Press, San Diego, CA, pp. 133–148.

26. Turk, E., Kerner, C. J., Lostao, M. P., and Wright, E. M. (1996) Membrane topology of the human $Na^+$/glucose cotransporter SGLT1. *J. Biol. Chem.* **271,** 1925–1934.

# II

## APPLICATIONS

# 7

# In Vitro Translation to Study HIV Protease Activity

**Zene Matsuda, Mutsunori Iga, Kosuke Miyauchi, Jun Komano, Kazuhiro Morishita, Akihiko Okayama, and Hirohito Tsubouchi**

## Summary

HIV-1 is an etiological agent of AIDS. One of the targets of the current anti-HIV-1 combination chemotherapy, called highly active antiretroviral therapy (HAART), is HIV-1 protease (PR), which is responsible for the processing of viral structural proteins and, therefore, essential for virus replication. Here, we describe an in vitro transcription/translation-based method of phenotyping HIV-1 PR. In this system, both substrate and PR for the assay can be prepared by in vitro transcription/translation. Protease activity is estimated by the cleavage of a substrate, as measured by enzyme-linked immunosorbent assay (ELISA). This assay is safe, rapid, and requires no special facility to be carried out. Our rapid phenotyping method of HIV-1 PR may help evaluate drug resistance, useful when choosing an appropriate therapeutic regiment, and could potentially facilitate the discovery of new drugs effective against HIV-1 PR.

**Key Words:** HIV-1; protease; genotyping; phenotyping; drug resistance; ELISA; protease inhibitor; p24.

From: *Methods in Molecular Biology, vol. 375,*
*In Vitro Transcription and Translation Protocols, Second Edition*
Edited by: G. Grandi © Humana Press Inc., Totowa, NJ

# 1. Introduction

## 1.1. HIV-1 Infection and AIDS

HIV-1 is a causative agent of AIDS, a syndrome that is epidemic worldwide. HIV-1 destroys a host's immune system by infecting its CD4+ T cells and macrophages. The high variability of HIV-1's genome poses a significant challenge to the development of effective preventive and therapeutic interventions against HIV-1 infection. A combination chemotherapy, called highly active antiretroviral therapy (HAART), remains one of the most effective anti-HIV-1 strategies and has been proven to improve the prognosis of HIV-1–infected individuals *(1,2)*.

HAART uses a mixture of drugs that individually and in combination target different aspects of virus replication. A major class of drugs used in HAART targets the viral enzymes essential for replication. One of the essential enzymes for HIV-1 replication is a protease encoded by the pol gene. This aspartyl protease is composed of a homodimer of a 99-amino acid-long monomer and responsible for the processing of the viral structural proteins Gag and Pol *(3)*. Without the proper processing by this protease, virions remain immature and noninfectious *(4)*. A drug-resistant protease (PR) has emerged and is rapidly becoming another challenge to anti-HIV-1 chemotherapy *(5,6)*. The adverse side effects of these protease inhibitors are also of great concern; an anti-HIV-1 chemotherapy regimen must be sustained over a prolonged period. The appropriate choice of PR inhibitors, therefore, is very important to proper management of therapies for HIV-1-infected individuals.

## 1.2. Genotyping vs Phenotyping

To achieve successful and economical chemotherapy, it is essential to select reagents that are appropriate for a particular patient. To do this, it is necessary to have information on the drug resistance of the virus infecting that patient *(7,8)*. There are two approaches for finding this information: genotyping and phenotyping the virus's HIV-1 genes of interest (e.g., pro) *(9,10)*.

Genotyping techniques that employ PCR and subsequent sequencing can handle many samples relatively easily. If an appropriate drug-resistance database is available, it can identify any previously known drug-resistant mutants. However, given the high mutation rates of HIV-1, it is often possible to encounter previously uncharacterized mutations, making it difficult to properly judge the virus's drug resistance.

Phenotyping provides a more direct measurement of the function of the gene of interest and offers crucial information on previously unidentified mutations. It is, however, usually more time consuming than genotyping, especially when using live viruses and dealing with multiple samples. For viral enzymes (including PR), a methodology that uses an in vitro biochemical assay is safer to perform than assays using intact live viruses, and it does not require use of a special biocontainment facility. In addition to evaluating drug resistance, phenotyping can be used to screen potential inhibitors against particular targets. Therefore, a relatively simple and rapid phenotyping methodology is highly desirable.

## 1.3. Phenotyping of HIV-1 Protease

Phenotyping of HIV-1 protease poses special difficulties. HIV-1 protease is produced from a precursor protein (Gag-Pol). The Gag-Pol precursor protein is generated by a -1 (minus 1) frame shift at the junction between gag and pol *(11)*. The protease portion is released from the precursor by the activated protease itself. Activation of protease activity requires a proper dimerization mediated through the preceding Gag portion of the precursor *(12)*.

The released mature HIV-1 PR characterized today usually starts with a proline residue. This proline residue contributes to the formation of the so-called flap region of the protease, and it is believed to be important for protease activity. An artificial expression of protease via direct attachment of a methionine residue in front of the proline residue may result in poor dimerization; this methionine residue also may negatively affect protease activity *(13)*. Furthermore, despite its specificity for the substrates, HIV-1 PR manifests a range of nonspecific cleavages to host proteins and, when over expressed, is toxic to the host. Expression of PR in a host as a fusion protein (nonfunctional form) may overcome this toxicity issue *(14)*, but the subsequent purification and refolding steps prohibit the processing of multiple samples simultaneously. Because HIV-1 PR is relatively small and requires proper dimerization to function, the attachment of a purification tag may not be desirable. These difficulties prompted us to apply an in vitro translation system to generate active protease for rapid phenotyping.

## 1.4. Design of the PR Phenotyping Using In Vitro Transcription and Translation

When we designed the system for the phenotyping of HIV-1 PR *(15)*, we tried to make it as simple and rapid a system as possible. We eliminated procedures and steps that would take a long time to finish, such as steps requiring extensive purification. The system is adaptable to measuring the activity of multiple proteases derived from different field isolates. The assay outline is shown in **Fig. 1**. The protease and substrate are generated first by a coupled, in vitro transcription/translation using a commercially available kit. The pro portion is fused to a preceding gag gene in-frame, making the expression of pro is no longer dependent on the relatively inefficient frame shifting. The amount of both the substrate and the protease can be estimated using an enzyme-linked immunosorbent assay (ELISA) method that measures the amount of the coexpressed surrogate protein (p24 in this system) (step 1). After protease digestion (step 2), the efficiency of the cleavage (i.e., the activity of the protease) is estimated by measuring the amount of uncleaved substrates using an ELISA method. This involves trapping the undigested substrates (step 3) and detecting the trapped substrates (step 4) by ELISA. All these steps make the method rapid and simple and possibly adaptable to high-throughput analysis of the PR activity.

## 2. Materials

### 2.1. Preparation and Quantitation of the Substrate and the Protease

1. TNT® Quick Coupled Transcription/Translation System (Promega, Madison, WI).
2. p24 ELISA kit RETRO-TEK® (ZeptoMetrix, Buffalo, NY).

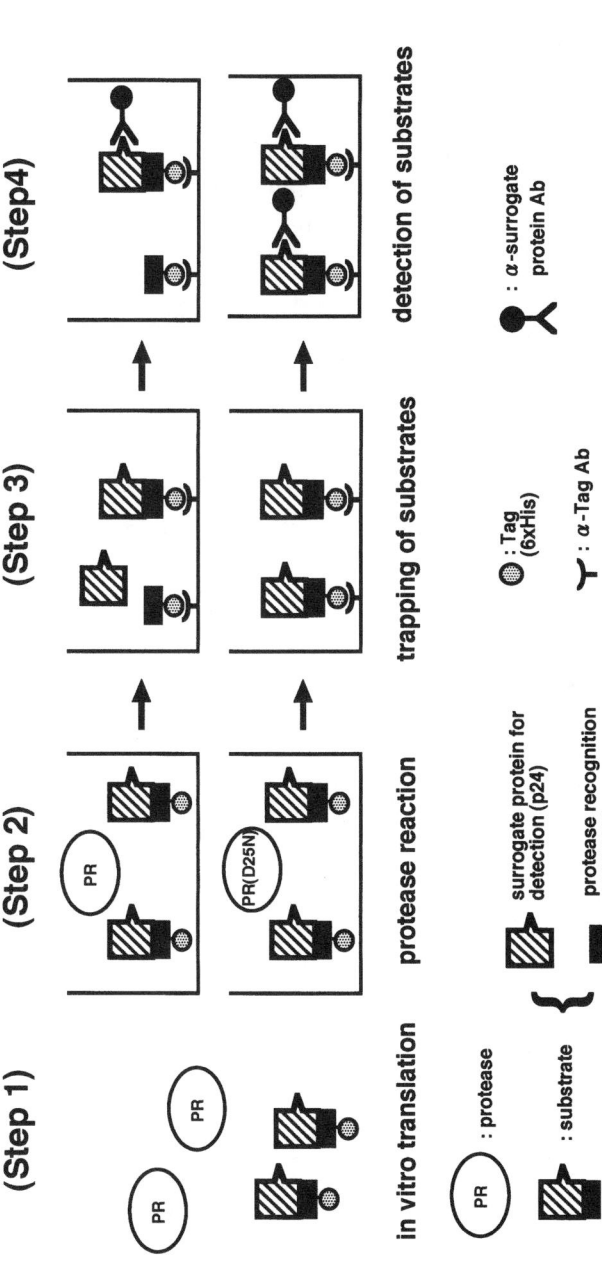

Fig. 1. An outline of the assay. The substrate and protease to be used are generated by a coupled in vitro transcription/translation (step 1). After quantitation of both the substrate and protease (not shown), both are mixed to initiate protease reaction (step 2). After digestion, the uncleaved substrate is trapped by the tag portion of the substrate (step 3) and the amount of the trapped substrate is quantitated using an ELISA method. (Modified from **ref. 15** with permission from Elsevier.)

## 2.2. Detection of the Uncleaved Substrate

1. MaxiSorp plates (Nalge Nunc International, Rochester, NY).
2. Anti-penta His monoclonal antibody (Qiagen, Hilden, Germany).
3. Coating buffer: 50 m$M$ sodium carbonate, pH 10.6.
4. Blocking buffer: 2.0% sucrose, 0.1% bovine serum albumin, 0.9% sodium chloride.
5. 4 methylumbelliferyl-β-D-galactoside (Sigma, St. Louis, MO).
6. Anti-p24 Fab-β-D-galactosidase conjugate *(16)*.

## 2.3. Chemicals

1. All chemical reagents are analytical grade and obtained from Sigma.
2. HIV-1 protease inhibitors can be obtained through the AIDS Research and Reference Reagent Program (US NIAID).

## 2.4. Equipment (Any Equivalent Machines Are Acceptable)

1. ELISA plate washer (Wellwash AC, Thermo Labsystems, Helsinki, Finland).
2. ELISA plate reader (Multiskan EX, Thermo Labsystems).
3. Incubator for 30 and 37°C.
4. Fluorometer (Fluoroskan II, Thermo Labsystems).

## 3. Methods

This method consists of four steps: (1) preparation of the substrate and protease by in vitro transcription/translation; (2) processing the substrate using protease; (3) trapping the unprocessed substrate with anti-His antibody; (4) quantitation of the trapped substrate by the ultrasensitive p24 ELISA (**Fig. 1**).

It is recommended that an anti-His monoclonal antibody-coated 96-well plate be prepared before starting the in vitro translation reaction (*see* **Subheading 3.3.1.**).

## 3.1. Preparation and Quantitation of the Substrates and Proteases by an In Vitro-Coupled Transcription/Translation

### 3.1.1. Preparation of Substrates and Proteases

The substrate and the protease used in this assay are made by in vitro transcription/translation using a rabbit reticulocyte-based TNT® Quick-coupled transcription/translation system according to the procedure suggested by the manufacturer (*see* **Note 1**). The DNA constructs used in this reaction are shown in **Fig. 2** (*see* **Note 2**). The substrate expression vectors are 5Zf(–)Gag-His and 5Zf(–)Gag-Pol-His. The protease expression vectors are 5Zf(–)PR(wt) and 5Zf(–)PR(D25N). Lysates programmed by 5Zf(–)PR(D25N) should be prepared simultaneously as a negative control group (*see* **Notes 3** and **4**).

### 3.1.2. The Quantitation of the Substrate and Protease by p24 ELISA (see **Note 5**)

The amount of substrate and protease generated by in vitro transcription/translation was estimated by measuring the amount of coexpressed p24 (*see* **Note 5**). This was done using a commercially available p24 ELISA kit (p24 ELISA kit RETRO-TEK®) according to the procedure suggested by its manufacturer (*see* **Note 6**).

## 3.2. Processing of the Substrate by Protease

1. The protease lysates programmed by 5Zf(–)PR(wt) and 5Zf(–)PR(D25N) by in vitro translation were mixed, respectively, with the substrate lysate programmed by 5Zf(–)Gag-His (or alternatively, 5Zf[–]Gag-Pol-His) to initiate the protease reaction. Typically in a total volume of 30 μL, 0.5–1 ng of p24-equivalent substrate and protease, respectively, were used. The mixture was incubated at 37°C for an appropriate time (*see* **Note 7**). When testing the activity of PR under the presence of the protease inhibitor, the inhibitor should be added to the reaction mixture (*see* **Note 8**). The reaction was set using a 96-well plate.
2. After incubation, the reaction was mixed with 200 μL of ice-cold PBS containing 0.2% BSA and transferred to the anti-His-coated plate (preparation of this plate is described in **Subheading 3.3.1.**).

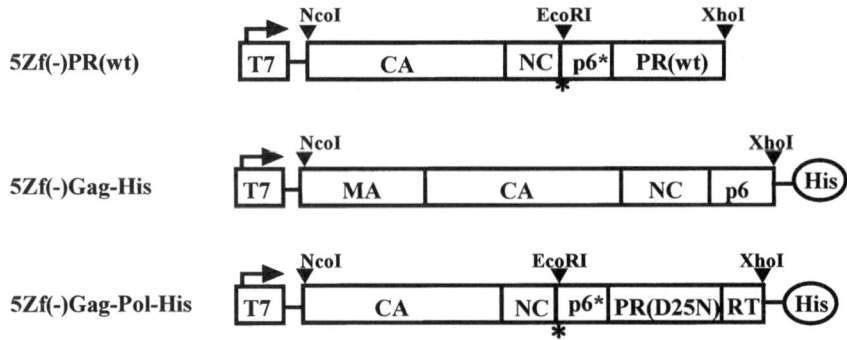

Fig. 2. DNA constructs used for in vitro transcription/translation. The protease expression vector expresses PR with the preceding Gag protein. The PR portion can be replaced with other PR, such as an EcoRI-XhoI fragment. At the EcoRI site, a forced frame shift is introduced to express PR in-frame with the preceding Gag protein. Forcing the frame shift removes dependence on a spontaneous frame shift, which usually occurs at about 5% efficiency. The two types of the substrate expression vectors 5Zf(–)Gag-His and 5Zf(–)Gag-Pol-His are shown. The latter lacks the MA portion but includes the PR and RT portions. There is a tag region (6xHis) that tethers the C-terminus of the substrate to the assay plate (*see* **Subheading 3.3.**). Both protease and substrate expression vectors have a T7 promoter to drive the expression. Matrix (MA), capsid (CA), nucleocapsid (NC), and P6 portion read in a pol frame (p6*) represent the components of the Gag protein. Other components include protease (PR), reverse transcriptase (RT), T7 promoter (T7), and 6x Histidine tag (6xHis), * the frame-shift site. The sites for the restriction enzymes, NcoI, EcoRI, XhoI, are shown. (Adopted and modified from **ref. 15** with permission from Elsevier.)

## 3.3. Measurement of Unprocessed Substrate by p24 ELISA

### 3.3.1. Preparation of Anti-His Monoclonal Antibody-Coated Plates

1. A MaxiSorp plate (Nalge Nunc International) was treated with 200 µL of coating buffer (50 mM sodium carbonate, pH 10.6) containing 5 mg/mL anti-penta His monoclonal antibody (Qiagen) at 4°C overnight. After incubation, the plate was washed once with PBS. The plate was then treated with blocking buffer (2.0% sucrose, 0.1% bovine serum albumin, 0.9% sodium chloride) at room temperature for 2 h (*see* **Note 9**).

2. The total protease reaction mixture was transferred into each well of the coated plate and incubated at 4°C overnight.
3. Measurement of the trapped unprocessed substrate using the ultrasensitive p24 ELISA (*see* **Note 10**).

## 3.4. Interpretation of the Data

In this assay, the activity of the protease (PR) was reflected by the reduction of optical density (OD) in the p24 assay. Because the extent of OD reduction is proportional to the protease activity, the activity of protease in an unknown sample (X) can be calculated using in the following formula.

% activity of PR = 100 × [OD PR(D25N) – OD PR (X)] / [OD PR(D25N) – OD PR(wt)].

In this formula, D25N is a null mutant of PR and wt represents a wild type PR.

## 3.5. Evaluation of the Drug-Resistant PR

As an example of how to apply the current method to an analysis for drug-resistant PR, we tested three drug-resistant PRs using this assay. The mutants tested were D30N, V82F, and L90M (for D30N, the original aspartic acid residue at the 30th position of PR was replaced with an asparagine residue). D30N and L90M are associated with resistance to nelfinavir and V82F is associated with resistance to indinavir (*17,18*). The activity of each of these mutants was compared with that of the wild type. We tested each mutant with two different substrates, one programmed with 5Zf(–)Gag-His and one programmed with 5Zf(–)Gag-Pol-His. The data are shown in **Fig. 3**. In the absence of the drug, all the mutants exhibited impaired PR activity compared with the activity in wild type. This result is consistent with the fact that most drug-resistant mutants gain resistance at the cost of losing enzymatic activity. This loss of fitness is a well-known phenotype of most currently known mutants (*19*).

It is noteworthy that there is a significant difference in the apparent protease activity in the two different substrates. Data on the Gag-Pol-His

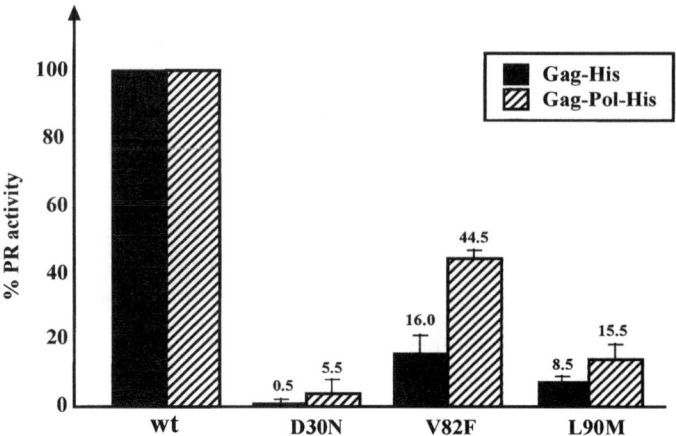

Fig. 3. Relative activity on different substrates for protease carrying mutations resistant to protease inhibitor. The activity of each mutant was evaluated on 5Zf(–)Gag-His or 5Zf(–)Gag-Pol-His substrates. For each mutant, the 5Zf(–)Gag-Pol-His substrate showed higher PR activity. The results are an average of three independent experiments. (Modified from **ref.** *15* with permission from Elsevier.)

substrate indicated higher activity. Using the Gag-Pol-His substrate, we compared activity of the wild type and of V82F against two different protease inhibitors, indinavir, and nelfinavir (**Fig. 4**). The V82F mutant showed 7–13 times greater resistance to indinavir when compared with wild type (0.2 µ*M* for wt vs 1.4–2.6 µ*M* for V82F) and 1.1 to 1.2 times greater resistance against nelfinavir (0.7 µ*M* for wt vs 0.8–1.5 µ*M* for V82F). Because V82F corresponds to one of the mutations associated with indinavir, the selective resistance of V82F against indinavir shown in this assay further verifies its relevance. The assay using actual viruses shows the same tendency *(17,18)*. The actual value of IC50, however, in this assay is usually much higher than the value obtained in the virus-based assay (*see* **Note 11**) *(20)*.

## 4. Notes

1. We found the wheat germ system did not work well, although we have not investigated the reasons for this. Typically we obtained about 5 ng of p24-equivalent proteins by using 0.5 mg of DNA in a 50-µL reaction incubated at 30°C for 90 min.

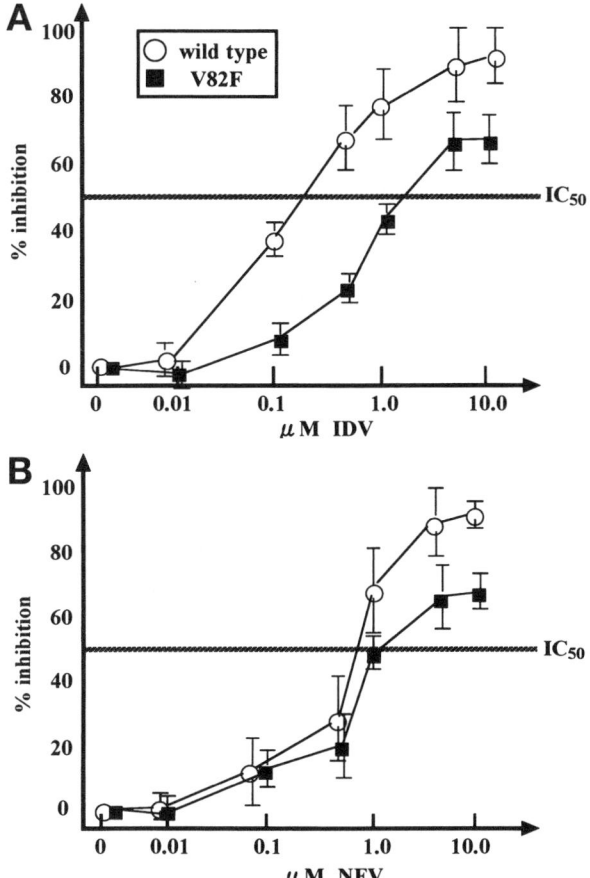

Fig. 4. Determination of IC50 of proteases. The wild type protease and a mutant protease, V82F, were tested for activity with indinavir (IDV) **(A)** and nelfinavir (NFV) **(B)**. Representative results for each are shown. For the wild type, the IC50 for IDV and NFV was 0.2 and 0.7 $\mu M$, respectively. The IC50 values for the V82F mutant for IDV and NFV were 1.4–2.6 $\mu M$ and 0.8–1.5 $\mu M$, respectively. (Modified from **ref. *15*** with permission from Elsevier.)

2. The DNA used here was prepared using a Qiagen Plasmid Maxi kit (Qiagen), but any similar procedure should be sufficient to provide DNA suitable for this purpose.
3. The 5Zf(–)Gag-His substrate vector contains the entire gag gene of the HIV-1 strain, HXB2, downstream of the T7 promoter. It also contains a 6xHis tag in-frame with the preceding *gag* gene. Another substrate vector,

5Zf(–)Gag-Pol-His, contains a portion of the *gag* gene (i.e., it is missing the MA portion) as well as a portion of the *pol* gene (corresponding to PR and part of RT) in front of the His tag. We found the 5Zf(–)Gag-Pol-His construct showed higher protease activity (2- to 10-fold) compared with that shown by the 5Zf(–)Gag-His substrate (**Fig. 3**). This may be a result of the difference in the target sequences or to the components incorporated into the construct, however, the exact reason for this is as yet unknown.

Essential components of the substrate are a protease recognition sequence, a surrogate protein marker to determine the amount of the substrate, and a tag sequence that tethers the substrate to the plate during the ELISA procedure. The target sequence is placed between the surrogate protein marker and the tag sequence. The target sequence can be short (around 10 amino acid residues), but it should be placed in a manner that allows the sequence to be accessible to the PR. The surrogate protein marker used for the quantitation can be any protein. Here, we employed p24 (*see* **Note 5**). The protease expression vector contains the *pro* gene preceded by the MA-missing portion of the *gag* gene. The 5Zf(–)PR(wt) and the 5Zf(–)PR(D25N) express an active and inactive protease, respectively, upon in vitro transcription/translation. The latter has the mutation at the active site of the protease (D25N); this inactivates the protease. In both constructs, pro was placed in-frame to the preceding gag sequence, therefore the expression of PR is not dependent on the –1 frame shift observed in natural HIV-1.

4. The *pro* region can be replaced with different *pro* DNA fragments as EcoRI-XhoI fragments. The preceding *gag* sequence is incorporated for the following two reasons.

   The nucleocapsid portion or upstream region has been reported to enhance the generation of an active dimer-form protease *(12)*. We failed to express active protease with simple addition of a methionine codon in front of the pro sequence (unpublished observation).

   With this configuration, the amount of the expressed protease can be estimated indirectly by measuring the amount of preceding coexpressed capsid (p24) protein (*see* **Note 5**).

5. Direct measurement of the protease using an ELISA method may seem more appropriate. However, if the goal is to assay the amount of several different mutant proteases, it is necessary to use a good antibody that can universally recognize diverse PRs with equal efficiency. Without such an antibody, the different immunoreactivities of the individual mutant proteases introduced into the construct could cause inaccuracies. Here we

used p24 as a surrogate protein marker to quantitate the cotranslated PR. We used p24 because there are several commercially available kits available that measure it. One can exchange the p24-portion with other surrogate markers, for example, with certain enzymes. If this is done, however, care should be taken to avoid choosing an enzyme that can be destroyed by the protease being expressed. The p24 protein is suitable because it will not be processed further by HIV-1 protease.

6. The prepared lysates can be stored at –80°C for later use. Although we did not have a need to test their durability to prolonged storage, we did find they remained durable when stored for a couple of days to a week.

7. The time of incubation needs to be calibrated. In our experience, when 67.5 ng/mL PR was used, the reaction proceeded linearly for as long as 60 min of incubation. We used a 45-min incubation period for our assay. The length of the incubation period may be affected by the amount of PR as well as by the substrate used in the assay. We observed an almost linear dose-dependence between 45 and 67.5 ng/mL p24-equivalent-PR in our assay *(15)*.

8. When evaluating the effect of an inhibitor, the stock solution of the inhibitor should be made in high concentration to reduce the amount of the organic solvent introduced into the reaction; inhibitors are hydrophobic and usually dissolve in organic solvents such as DMSO or ethanol. We did not adjust the pH of the reaction mixture to a lower pH that is known as an optimum for PR. Even at neutral pH, the reaction proceeds.

9. In our experience, the anti-His monoclonal antibody-mediated trapping had a better binding activity than direct trapping using Ni- or another metal-based binding.

10. The ultrasensitive p24 assay was used for this purpose. The details of the assay are described by Hashida et al. *(16)*. Briefly, the unprocessed substrate trapped on anti-His monoclonal antibody was incubated with 25 p$M$ monoclonal mouse anti-p24 Fab-β-D-galactosidase conjugate at room temperature for 1 h. After incubation, the well was washed and the bound β-D-galactosidase activity was assayed at 30°C for 1 h by fluorometry using 4 methylumbelliferyl-β-D-galactoside as a substrate. The fluorescence intensity is measured relative to that of $1 \times 10^{-8}$ mol/L of 4-methylumbelliferone. The use of an ordinary p24 assay kit may result in higher background.

11. The actual value of IC50 obtained in our assay was usually much higher than the value obtained in the method using replicating viruses in a cell-culture system. Originally we reasoned that this was partly attributable to the milieu of various proteins in the in vitro transcription/translation system. However, Pettit et al. have recently shown that the PR produced in in

vitro transcription/translation is less sensitive to inhibitors compared to the mature PR *(20)*. This immature nature of PR may be responsible for the observed high IC50 in our system.

# References

1. Besch, C. L. (2004) Antiretroviral therapy in drug-naive patients infected with human immunodeficiency virus. *Am. J. Med. Sci.* **328,** 3–9.
2. Yeni, P. G., Hammer, S. M., Hirsch, M. S., et al. (2004) Treatment for adult HIV infection: 2004 recommendations of the International AIDS Society-USA Panel. *JAMA* **292,** 251–265.
3. Meek, T. D., Dayton, B. D., Metcalf, B. W., et al. (1989) Human immunodeficiency virus 1 protease expressed in *Escherichia coli* behaves as a dimeric aspartic protease. *Proc. Natl. Acad. Sci. USA* **86,** 1841–1845.
4. Gottlinger, H. G., Sodroski, J. G., and Haseltine, W. A. (1989) Role of capsid precursor processing and myristoylation in morphogenesis and infectivity of human immunodeficiency virus type 1. *Proc. Natl. Acad. Sci. USA* **86,** 5781–5785.
5. Johnson, V. A., Brun-Vezinet, F., Clotet, B., et al. (2004) Update of the drug resistance mutations in HIV-1: 2004. *Top. HIV Med.* **12,** 119–124.
6. Kutilek, V. D., Sheeter, D. A., Elder, J. H., and Torbett, B. E. (2003) Is resistance futile? *Curr. Drug Targets Infect. Disord.* **3,** 295–309.
7. Cohen, C. J., Hunt, S., Sension, M., et al. (2002) A randomized trial assessing the impact of phenotypic resistance testing on antiretroviral therapy. Aids **16,** 579–588.
8. Durant, J., Clevenbergh, P., Halfon, P., et al. (1999) Drug-resistance genotyping in HIV-1 therapy: the VIRADAPT randomised controlled trial. *Lancet* **353,** 2195–2199.
9. Zhang, M. and Versalovic, J. (2002) HIV update. Diagnostic tests and markers of disease progression and response to therapy. *Am. J. Clin. Pathol.* **118,** S26–S32.
10. Maldarelli, F. (2003) HIV-1 fitness and replication capacity: what are they and can they help in patient management? *Curr. Infect. Dis. Rep.* **5,** 77–84.
11. Jacks, T., Power, M. D., Masiarz, F. R., Luciw, P. A., Barr, P. J., and Varmus, H. E. (1988) Characterization of ribosomal frameshifting in HIV-1 gag-pol expression. *Nature* **331,** 280–283.
12. Zybarth, G. and Carter, C. (1995) Domains upstream of the protease (PR) in human immunodeficiency virus type 1 Gag-Pol influence PR autoprocessing. *J. Virol.* **69,** 3878–3884.

13. Ishima, R., Torchia, D. A., Lynch, S. M., Gronenborn, A. M., and Louis, J. M. (2003) Solution structure of the mature HIV-1 protease monomer: insight into the tertiary fold and stability of a precursor. *J. Biol. Chem.* **278,** 43,311–43,319.

14. Rangwala, S. H., Finn, R. F., Smith, C. E., et al. (1992) High-level production of active HIV-1 protease in *Escherichia coli. Gene* **122,** 263–269.

15. Iga, M., Matsuda, Z., Okayama, A., et al. (2002) Rapid phenotypic assay for human immunodeficiency virus type 1 protease using in vitro translation. *J. Virol. Methods* **106,** 25–37.

16. Hashida, S., Hashinaka, K., Nishikata, I., et al. (1996) Ultrasensitive and more specific enzyme immunoassay (immune complex transfer enzyme immunoassay) for p24 antigen of HIV-1 in serum using affinity-purified rabbit anti-p24 Fab' and monoclonal mouse anti-p24 Fab'. *J. Clin. Lab. Anal.* **10,** 302–307.

17. Patick, A. K., Duran, M., Cao, Y., et al. (1998) Genotypic and phenotypic characterization of human immunodeficiency virus type 1 variants isolated from patients treated with the protease inhibitor nelfinavir. *Antimicrob. Agents Chemother.* **42,** 2637–2644.

18. Condra, J. H., Holder, D. J., Schleif, W. A., et al. (1996) Genetic correlates of in vivo viral resistance to indinavir, a human immunodeficiency virus type 1 protease inhibitor. *J. Virol.* **70,** 8270–8276.

19. Clementi, M. (2004) Can modulation of viral fitness represent a target for anti-HIV-1 strategies? *New Microbiol.* **27,** 207–214.

20. Pettit, S. C., Everitt, L. E., Choudhury, S., Dunn, B. M., and Kaplan, A. H. (2004) Initial cleavage of the human immunodeficiency virus type 1 GagPol precursor by its activated protease occurs by an intramolecular mechanism. *J. Virol.* **78,** 8477–8485.

# 8

# The Protein Truncation Test in Mutation Detection and Molecular Diagnosis

## Oliver Hauss and Oliver Müller

## Summary

The protein truncation test (PTT) is a simple and fast method to screen for biologically relevant gene mutations. The method is based on the size analysis of products resulting from in vitro transcription and translation. Proteins of lower mass than the expected full-length protein represent translation products derived from truncating frame shift or stop mutations in the analyzed gene. Because of the low sensitivity of the conventional PTT mutations can be detected only in those samples, which harbor a high relative number of mutated gene copies. This disadvantage can be overcome by technical modifications and advanced forms of the PTT. Modifications like gene capturing and the digital PTT lower the detection limit and thus allow the use of the PTT in the detection of mutations in body fluids. Another disadvantage of the conventional PTT is the use of radioactive labels for protein detection. Recently, modifications like fluorescent labels or the use of tagged epitopes were established, which allow the detection of the nonradioactive translation product. When several epitopes in different reading frames are used, the mutation detection spectrum can be expanded to all possible frame shift mutations. These modifications transform the PTT into a powerful nonradioactive technique to detect mutations with high sensitivity.

**Key Words:** Protein truncation test; PTT; in vitro-synthesized protein assay; IVSP; mutation detection; early tumor diagnosis.

From: *Methods in Molecular Biology,* vol. 375,
*In Vitro Transcription and Translation Protocols, Second Edition*
Edited by: G. Grandi © Humana Press Inc., Totowa, NJ

# 1. Introduction

## 1.1. The Conventional Protein Truncation Test

More than 10 yr ago the in vitro transcription and translation was introduced as a novel method in mutation detection *(1–4)*. The protein truncation test (PTT), also known as the in vitro-synthesized protein assay (IVSP), is most suited for the scanning of kilobase-sized DNA fragments for mutations like frame shift or stop mutations that lead to premature translation termination (for reviews *see* **refs. 5** and **6**). The conventional PTT can be divided into four stages: the PCR of the genomic exon fragment or the reverse transcription-PCR of the mRNA, the in vitro transcription and translation in the presence of [$^{35}$S] labeled methionine or [$^{14}$C] labeled leucine, the sodium dodecyl sulfate-polyacrylamide gel electrophoresis (SDS-PAGE) of the translated proteins and the autoradiographic detection of the labeled proteins. Proteins of lower mass than the expected full-length protein represent translation products derived from truncating frame shift or stop gene mutations.

The PTT assay allows the fast screening for relevant mutations in large coding segments of tumor suppressor genes or genes responsible for inherited diseases in a single assay. For example, 35 PCR fragments are required to screen the adenomatous polyposis coli *(APC)* gene by denaturing gradient gel electrophoresis, whereas just five are required using the PTT *(6)*. The method finds application in the diagnostic screening for inherited or somatic mutations in many different genes including *APC* (accession number in the NCBI online database Mendelian inheritance in man (OMIM) (http://www.ncbi.nlm.nih.gov): OMIM175100), BRCA1 and 2 (breast cancer susceptibility gene; OMIM113705 and OMIM600185), cystic fibrosis transmembrane conductance regulator (OMIM602421), and tuberous sclerosis complex (OMIM191092). The PTT was modified and adapted to optimize the analysis of several specific genes (for a list of gene-specific technical modifications *see* **ref. 6**). Besides its many advantages the conventional PTT holds several technical obstacles, which prevent its broad clinical application. These are (1) the necessity of radioactive labels and the costs for laboratory infrastructure and waste disposal, (2) the time-consuming blotting procedure, and (3) the low sensitivity. These restrictions make the conventional PTT unusable in high-throughput screening and in the diagnostic analysis of samples with low copy numbers of mutated alleles.

## 1.2. The Advanced PTT

Technical modifications and further developments of the PTT include the labeling of the translated protein by fluorescent or biotinylated amino acids or by the introduction of immunodetectable tags to circumvent the use of radioactive labels. Other technical modifications are the gene capture and the digital PTT with the aim to increase the mutation detection sensitivity. Until now several advanced forms of PTT have been established. These PTTs are based on the combined application of these modifications and have the potential to be used in high-sensitive, nonradioactive high-throughput mutation detection.

### 1.2.1. Digital PTT

The digital PTT was developed to increase the mutation detection sensitivity *(7)*. The most important step is the distribution of the ready-made PCR mix containing the original DNA template to a high number of small aliquots and the parallel PCR and analysis of the aliquots. Because of the dilution many aliquots do not contain any mutant gene copy, whereas the relation of the number of mutated copies to wild type copies is increased in a few aliquots. After a second PCR followed by the transcription and translation in the presence of [$^{35}$S] labeled methionine, the resulting proteins are analyzed by SDS-PAGE and autoradiography. Digital PTT preceded by a gene capture step was used successfully to detect *APC* mutations in fecal DNA samples from patients with early stages of colorectal tumors *(7)*. The major disadvantage of digital PTT is the number of aliquots to be analyzed. Thus, the simultaneous analysis of several patient samples requires automation.

### 1.2.2. Fluorescent Labeling

To avoid the use of radioactive labels, the proteins can be labeled with fluorescent amino acids *(8)*. The lys-tRNA, which is charged with fluorescently labeled lysine, is added to the in vitro translation mix, and fluorescent lysine residues are incorporated by the reaction. The detection of the labeled proteins is accomplished directly "in-gel" by the use of a laser-based fluorescent gel scanner. This technique makes the time-consuming blotting obsolete, which is otherwise necessary when the proteins are labeled by biotin or by immunodetectable tags. The fluorescent protein labeling was further enhanced by the use of multicolor labels and combined with an improved form of digital PTT and tag fusion *(9)*. This multicolor in vitro translation consists of several different steps. After the enrichment of the gene by hybrid capture the gene copy number is quantified. A defined gene copy number is distributed into several reaction wells and the gene is amplified by PCR. The PCR product is in vitro transcribed and translated in the presence of one of four differently labeled charged lys-tRNAs, which yield four correspondingly labeled translation products. Four differently colored aliquots resulting from one sample are pooled. The full length "false-positive" proteins are removed with an antibody against a carboxyterminal tag, which was introduced in the PCR step. Detection is done by SDS-PAGE and laser scanning. The fluorescent and the multicolor PTT require tremendous investments in reagents (fluorescently labeled lys-tRNAs) and technical equipment (laser-based fluorescent gel scanner, pipetting workstation). Nevertheless because the method is fast and highly sensitive, it may find several applications in the fields of research and diagnosis.

### 1.2.3. Tagging

Recently, the addition of aminoterminal tags to the in vitro translated proteins by tag-coding primers in the preceding PCR was established *(10)*. This modification allows the nonradioactive and selective detection of the correctly initiated translation products by Western blot analysis *(10,11)*. Thereby, the number of false-positives resulting from secondary translation initiation is decreased significantly. Aminoterminal tags have also been used in the positive selection of the translated protein *(9,11,12)*. The alternative or the additional introduction of one or more carboxyterminal tags might also be advantageous. A tag at this end might serve the negative selection of unwanted full-length proteins or the immunodetection of the protein *(9,11)*. General examples for possible tags together with the corresponding detection reagents are listed in **Table 1**.

**Table 1**
**Examples for Tags Suitable for Protein Detection**

| Tag | Peptide sequence | DNA coding sequence (5'-3') | Binding and detection reagent |
|---|---|---|---|
| lys-biotin | (K-bio)5 | (AAA)5 | (strept)avidin |
| arg-biotin | (R-bio)5 | (CGA)5 | mAk, pAk |
| lys- digoxigenine | (K-dig)5 | (AAA)5 | |
| arg-digoxigenine | (R-dig)5 | (CGA)5 | |
| (RGS)His | (RGS)HHHHHH | (AGG GGC AGC) CAC CAC CAC CAC CAC CAC | Ni$^{2+}$ matrix, mAk |
| flag | DYKDDDDKG | GAC TAC AAG GAC GAC GAC GAC AAG GGT | mAk |
| glu-glu | EFMPME | GAG TTC ATG CCC ATG GAG | mAk |
| Hemagglutinin (HA) | YPYDVPDYA | TAC CCC TAC GAC GTG CCC GAC TAC GCC | mAk, pAk |
| KT3 | TPPPEPET | ACA CCA CCA CCA GAA CCA GAA ACA | mAk |
| c-myc | EQKLISEEDL | GAG CAG AAG CTG ATC AGC GAG GAG GAC CTG | mAk, pAk |
| protein C | EDQVDPRLIDGK | GAG GAC CAG GTG GAC CCC AGG CTT ATC GAC GGC AAG | mAk |
| Strep-tag | WSHPQFEK | TGG AGC CAC CCG CAG TTC GAA AAA | (strept)avidin, mAk |
| V5 | EGKPIPNPLLGLDST | GAG GGC AAG CCC ATC CCC AAC CCC CTG CTG GGC CTG GAC AGC ACC | mAk |
| vsv-G | YTDIEMNRLGK | TAC ACC GAC ATC GAG ATG AAC AGG CTG GGC AAG | mAk |

Examples of tags for protein labeling. In the cases of lys- or arg-biotin and lys- or arg-digoxigenine five lysines or five arginines are coded by the primers. The corresponding covalently modified amino acids as such or as charged to tRNAs are added to the in vitro translation reaction. All other tags comprise biogenic amino acids and are directly coded. The His-tag might be prolonged with the sequence RGS at the aminoterminal end to increase the detection specificity and sensitivity. Binding and detection of the tags are performed with monoclonal or polyclonal antibodies (mAk, pAk), or other binding partners with high affinity.

## 1.2.4. High-Sensitive PTT for the Detection of All Possible Stops and Frame Shifts

Recently, we introduced a modified PTT, which allows the high-sensitive nonradioactive detection of all possible stop and frame shift mutations *(11,12)*. We established the method for the analysis of the APC gene (for details *see* **Subheading 3.**). After DNA amplification, the gene of interest is enriched by binding to an immobilized oligonucleotide. In the following PCR a tag coding primer is used as aminoterminal primer. The fusion of two consecutive hemagglutinin (HA)-tags  at the aminoterminus might increase the overall yield of HA-tagged proteins and the average number of HA-tags per translated protein and thus the detection signal in the subsequent Western blot *(11)*.

In case frame shift mutations that do not result in premature stop codons should be detected, different tags are fused to the carboxyterminal end in different frames **(Fig. 1)**. The method was established by adding a protein C-tag in the translation frame of the amplified wild-type gene sequence (tag B in **Fig. 1**), and two repetitive V5-tags (tag C in **Fig. 1**), each of them in an alternative reading frame, to the carboxyterminal end.

Next the PCR product is used as template in a second PCR, which serves two aims. First, the amount of PCR product is increased, which is necessary for the following in vitro transcription reaction. Second, the addition of the coding sequences for the tag and the necessary motifs (i.e., promoter, Kozak motif, start codon) can be distributed on two different primers. This strategy shortens the necessary primers length.

The PCR fragment is transcribed and the resulting mRNA is translated in a one-tube transcription/translation reaction. As an optional step, the correctly initiated primers can be enriched (positive selection) and the full-length proteins can be removed from the sample (negative selection) by immunoprecipitation of the corresponding tags. These steps decrease the background caused by the wild-type protein *(9,11)*.

The detection of the tagged proteins is performed by conventional immunodetection after Western blot. We showed that the fusion of carboxyterminal tags in different translation frames allows the detection of frame shift mutations, which do not lead to stop signals. In our system, the analyzed frame shift mutations lead to the translation of the new V5-tag fused to the carboxyterminal end in one of the two reading frames alternative to the original frame. Thereby, the detection of a protein with the similar size as the wild type protein using anti-V5 antibodies indicated the presence of these mutations (**Fig. 1**). Using the *APC* gene fragment from nucleotide 3580 to 4260 as an example, a 1-bp insertion might not necessarily lead to a stop signal unless it is located more than 284 bp upstream of the 3'-end of the fragment. Thus, frame shift mutations in the 3'-region of the analyzed *APC* gene fragment might lead to false-negative results. The use of carboxyterminal fragments in all three possible reading frames makes these mutations analytically accessible. In addition, even mutations, which are located very close to the 3'-region of a gene fragment, are detected. Therefore, the analysis of extended overlapping consecutive fragments of a single exon is not necessary, when the analyzed gene region is too long for a one-step analysis.

## *1.3. Further Perspectives*

In the presented PTT method technical aspects of the hybrid capture technique and of the PTT assay are combined and modified. With these modifications the method allows the sensitive detection of stop mutations and also of virtually any possible frame shift mutation, no matter whether the mutation leads to a stop signal within the analyzed gene fragment. The method has been successfully used in the nonradioactive and sensitive detection of tumor relevant gene mutations *(11,12)*. In addition the analysis of amino- and carboxyterminally tagged proteins allows the further development of an enzyme-linked immunosorbent assay (ELISA)-based mutation detection method, which was proposed recently *(5,13)*. For this, the fusion of a second aminoterminal tag (tag A' in **Fig. 1**) is necessary. The in vitro translated protein is immobilized via the first aminoterminal tag (tag A). The concentrations of tag A' and of a carboxyterminal tag are quantified by ELISA. The resulting ratio between the second aminoterminal tag and the carboxyterminal tag would then allow the conclusion whether the sample contains mutated proteins or not. Such an ELISA could be easily adapted to a high-throughput system allowing the use of the PTT in diagnostic screening programs. The principle of the ELISA detection of tagged proteins was recently modified *(14)*. The protein was immobilized via a biotin, which was introduced in the in vitro translation. In addition optional fluorescent

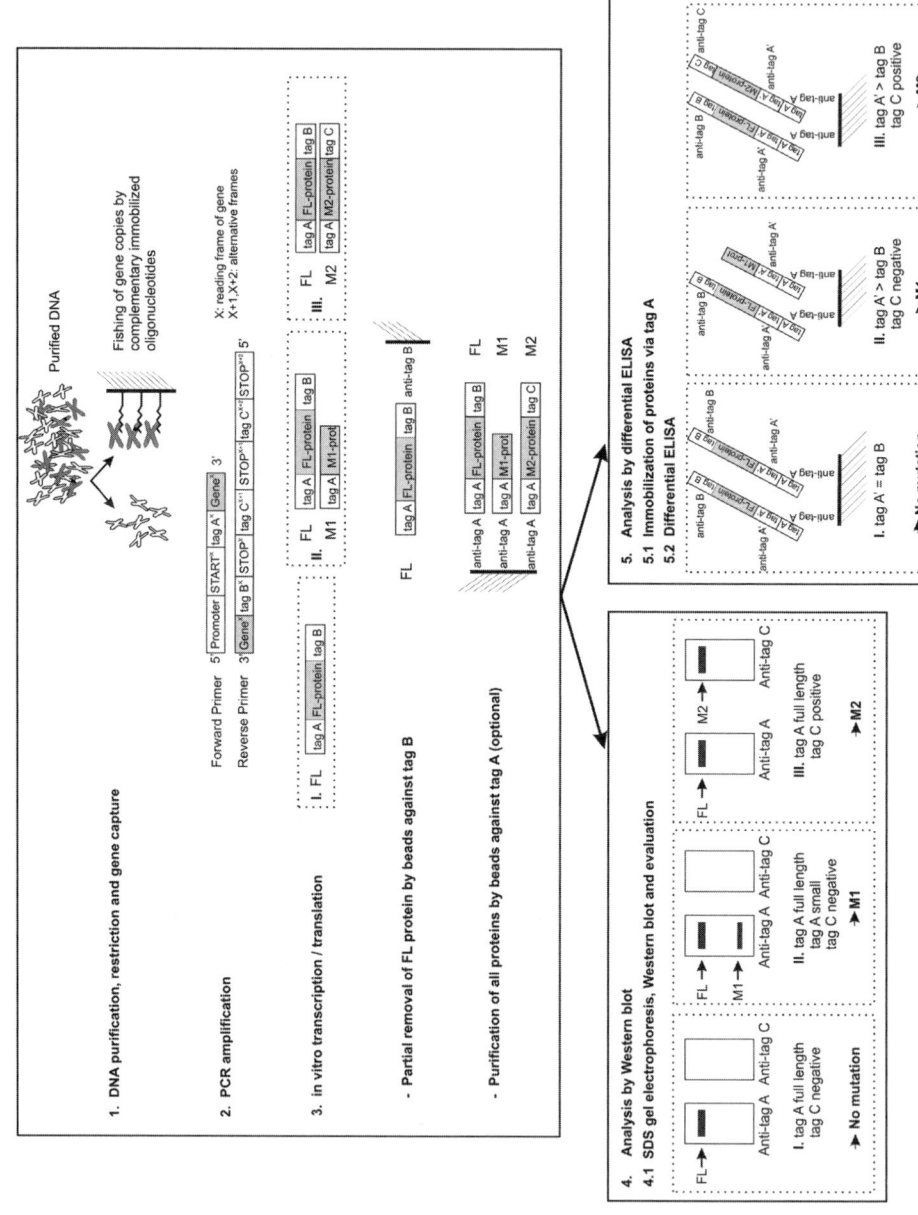

labels were proposed, which would allow the comparison of the ELISA results with the results obtained by SDS-PAGE and laser scan analysis.

Altogether, the PTT represents a powerful method for the sensitive detection of relevant frame shift and stop mutations. Nevertheless, there are still a lot of technical modifications and improvements necessary until this method will be used as a high-throughput method in routine mutation screening programs.

---

Fig. 1. *(opposite page)* Schematic overview of the method. After purification of the DNA the gene copies to be analyzed are captured by hybridization to immobilized complementary sequences (step 1). The gene fragment is amplified by PCR (step 2). The schematic structures of the primers to be used for the analysis of truncating and nontruncating frame shift mutations are shown. In case only truncating mutations should be detected, the carboxyterminal tags are not necessary. Exemplified primer sequences for the analysis of the APC gene are listed in **Table 2**. The following in vitro transcription and translation can yield in different proteins (step 3). These are the full-length protein (FL), which is synthesized owing to the presence of wild type gene in the sample, and one of the two truncated proteins M1 and M2. Whether M1 or M2 is synthesized depends on the position and the type of the mutation. To decrease the nonspecific background the full length protein is partly removed and all proteins, which start with the correct start codon, are enriched by binding to an immobilized matrix via the terminal tags. The protein(s) are detected by SDS-PAGE and differential Western blot (step 4). The absence of a band in the anti-tag A analysis below the full length protein together with the absence of any band in the anti-tag C analysis indicate the absence of any frame shift or stop mutation (left). The presence of a small band in the anti-tag A analysis indicates the presence of a mutation, which leads to a significantly shorter protein (middle). The presence of a protein band in the anti-tag C analysis indicates the presence of a mutation, which leads to a frame shift with no effect on the length of the translated protein (right). Alternatively, the proteins can be analyzed by differential ELISA (step 5). In this case the addition of the additional tag A' at the aminoterminus is necessary. The proteins are immobilized via tag A, and the tags A', B, and C are quantified. The detected relative amounts of the tags allow conclusions on the presence and the type of the mutations.

## 2. Materials

Kits for DNA purification:
1. QIAamp DNA Mini Kit or QIAamp DNA Stool Mini Kit (Qiagen).
2. DNA restriction: *Dra*I with buffer NEB4 (New England Biolabs).
3. Oligonucleotides: custom-synthesized by MWG Biotech; for sequences, *see* **Table 1**.
4. Reagents for gene capturing: magnetic beads coupled with streptavidine, magnetic racks, and B+W buffer (Dynal Biotech); wash 10 µL suspension of beads twice with 40 µL of 1X B+W buffer and resuspend in 100 µL 2X B+W buffer right before use.
5. PCR enzyme: Taq DNA polymerase (Qiagen).
6. In vitro transcription and translation: TNT T7 Quick for PCR DNA (Promega); add 5 µL of 1 m$M$ methionine and 20 µL of nuclease-free water (both provided in the kit) to 200 µL reaction mix; aliquot in 11.25 µL ready-mix batches in the tubes, which will be used for the reactions and store at –80°C until use.
7. Membrane: PVDF Hybond-P membrane (Amersham Biosciences); activate in methanol; blot at 5 mA/cm$^2$; for 30 min in a semidry blotting chamber weighted with cooling slabs.
8. Transfer buffer: 25 m$M$ Tris, 200 m$M$ glycine, 10% methanol.
9. Phosphate buffered saline (PBS): 150 m$M$ NaCl, 10 m$M$ NaH$_2$PO$_4$/Na$_2$HPO$_4$, pH 7.2.
10. HRP-conjugated anti-HA-antibody (Roche).
11. Detection Reagent: ECL Plus (Amersham Biosciences).
12. Film: Hyperfilm ECL (Amersham Biosciences).

## 3. Methods

### 3.1. DNA Purification, Restriction, and Gene Capture

1. Purify DNA from tumor tissue or stool using commercial reagents and following the protocols of the supplier.
2. For capturing of the APC gene from stool DNA digest the DNA with the restriction enzyme *Dra*I (*see* **Note 1**). To 100 µL DNA solution, add 11.4 µL buffer NEB4, and 3 µL (60 U) enzyme. Digest for 2 h at 37°C. Add 10 pmol biotin-labeled primer APC1 to 100 µL *Dra*I digested stool DNA. After denaturing and reannealing at 95°C (5 min) and 56°C (10 min), add

100 µL streptavidin-coupled magnetic beads and rotate over night at 4°C. Wash twice with 100 µL 1X B+W buffer and once with 500 µL water. Resuspend in 35 µL water. Use the DNA coupled to the precipitated beads directly in the following PCR.

## 3.2. PCR Amplification

Reactions are done in a total volume of 50 µL.

1. Amplify the mutation cluster region of the APC gene in a two-loop PCR with primers APC2 and APC3 (**Table 2**): 95°C (5 min) // [94°C (30 s) / 56°C (30 s) / 72°C (90 s)] 20x // [94°C (30 s) / 58°C (30 s) / 72°C (90 s)] 20X // 72°C (10 min). Final concentrations in the reaction are: BSA: 0.1 µg/µL, $MgCl_2$: 1.5 m$M$, dNTP: 0.2 m$M$, in the presence of 2.5 U Taq polymerase and 20 pmol of each primer (*see* **Note 2**). Analyze the PCR products on a 1% agarose gel (*see* **Note 3**).
2. For second PCR use 5 µL of the first reaction (*see* **Note 4**), the primer APC4 and the primer APC3 in the temperature profile: 95°C (5 min) // [94°C (30 s) / 60°C (30 s) / 72°C (90 s)] 40X // 72°C (10 min) in the presence of 1.5 m$M$ $MgCl_2$, 0.2 m$M$ dNTP, 20 pmol of each primer, and 5 U Taq polymerase. Repeat detection by agarose gel electrophoresis.

## 3.3. In Vitro Transcription and Translation

1. Add 1.25 µL of the product of the second PCR to ready-made 11.25 µL of the T7 in vitro transcription and translation kit, conduct the reaction at 30°C over 90 min (*see* **Note 5**).
2. To decrease the background caused by the wild-type protein, remove the full length protein (negative selection) and enrich the correctly initiated protein (positive selection) by immobilization using the corresponding antibodies coupled to magnetic beads (**Fig. 1**).

## 3.4. Analysis and Detection by Western Blot

1. Separate proteins on a 12% SDS-PAGE and electro-blot to PVDF membrane using standard procedures. Block membrane for 30 min with 5% nonfat dry milk in TPBS (PBS, 0.1% Tween-20) and incubate for 1 h with an horseradish peroxidase (HRP)-conjugated anti-HA antibody diluted 1:5000 in TPBS. After washing with TPBS, detect the signal by enhanced chemiluminescence (ECL) and fluorography. Protein products of lower sizes as compared to the full length control protein indicate the presence

**Table 2**
**Primers Used for the PTT Analysis of APC**

| Primer | Sequence | Annealing | Addition of | Used for |
|---|---|---|---|---|
| APC1 | biotin-TCCTTCATCACAGAAACAGT | 3618-3673 in APC gene | biotin | gene capture |
| APC2 | CGCCATGTACCCCTACGACGTGCCCGACTACGCCC AGAAACAGTCATTTTCATTCTCA | 3628-3651 in APC gene | start codon, HA-tag | 5' primer, first PCR |
| APC3 | CCGTCATTCATTCCCATTGTCATTTTCCT | 4665-4643 in APC gene | stop codon | 3'primer, first PCR, second PCR |
| APC4 | ATCCTAATACGACTCACTATAGGGAGCCACCATG TACCCCTACGACGTG | APC2 and APC5 | T7-promoter, kozak motif | 5' primer, second PCR |
| APC5 | CTACTTGCCGTCGATAAGCCTGGGGTCCACCTGGT CCTCTTTTTCTGCCTCTTTCTCTTGGTT | 4722-4699 in APC gene | protein C-tag, stop codon in wild type frame | 3'primer, first PCR |
| APC6 | CTAGGTGGCTGTCCAGGCCCAGCAGGGGGTTGGGG ATGGGCTTGCCCCTACTTGCCGTCGATAAGCC | APC5 | V5 tag, stop codon in +1 frame | 3'primer, second PCR |
| APC7 | CTAGGTGGCTGTCCAGGCCCAGCAGGGGGTTGGGG ATGGGCTTGCCCCCTACTTGCCGTCGATAAGCC | APC5 | V5 tag, stop codon in +2 frame | 3'primer, second PCR |

Primers used for mutation analysis of the APC gene by PTT. The primer pair APC2/APC3 is used for the first PCR and the primer pair APC4/APC3 for second PCR. In case that the nontruncating frame shift mutations should be detected, the primer pair APC2/APC5 is used for the first PCR and the primers APC4, APC6 and APC7 are used for the second PCR. The sequence motifs in primer sequences are coded as follows: APC, kozak, stop, start, HA-tag, T7-promoter, protein C-tag, V5-tag.

of truncation causing mutations. Step 4 in **Fig. 1** shows the differential Western blot analysis of both truncating and nontruncating frame shift mutations. When only truncating mutations should be detected, only one Western blot is necessary.

## 4. Notes

1. This has been found to increase the yield in the following hybrid capture step for APC. For other genes, other restriction enzymes may be suitable. This restriction and the gene capture step are only necessary for samples with low level of alleles to be analyzed and can be skipped when directly working with tumor DNA.
2. In case that all possible frame shift mutations should be detected, note the different primer sets (**Table 2**, **Fig. 1**): Use primers APC2 and APC5 for the first PCR and APC4 together with a 1:1 mixture of APC6 and APC7 for the second PCR. By this strategy also proteins with mutations, which do not lead to premature stops and thus to a truncated protein, are detectable by the presence of a V5 tag (**Fig. 1**). Certainly, three different tags could be used at the 3'-end. This would even allow differentiating between all three reading frames. For examples of possible protein tags *see* **Table 1**.
3. Usually, especially when working with stool samples, only the positive control is detectable at this step.
4. It has been found to be beneficial to do two reactions with differing volume of captured DNA solution in the PCR process. Using more DNA in the amplification sometimes yields less, not more product. The authors split the 35 µL volume yielded in the gene capturing step into PCR reactions with 5 or 30 µL templates, respectively.
5. Make sure to cool down the reaction to 4°C immediately after the reaction is finished and, if not proceeding to the next step, freeze to –20°C as soon as possible.

## References

1. Powell, S. M., Petersen, G. M., Krush, A. J., et al. (1993) Molecular diagnosis of familial adenomatous polyposis. *N. Engl. J. Med.* **329,** 1982–1987.
2. Roest, P. A., Roberts, R. G., Sugino, S., et al. (1993) Protein truncation test (PTT) for rapid detection of translation- terminating mutations. *Hum. Mol. Genet.* **2,** 1719–1721.

3. Kinzler, K. W. and Vogelstein, B. (1993) Molecular diagnosis of familial adenomatous polyposis. US patent no. 5,709,998.
4. van der, L. R., Khan, P. M., Vasen, H., et al. (1994) Rapid detection of translation-terminating mutations at the adenomatous polyposis coli (APC) gene by direct protein truncation test. *Genomics* **20,** 1–4.
5. Den Dunnen, J. T. and Van Ommen, G. J. (1999) The protein truncation test: a review. *Hum. Mutat.* **14,** 95–102.
6. Wallis, Y. (2004) Mutation scanning for the clinical laboratory-protein truncation test. *Methods Mol. Med.* **92,** 67–79.
7. Traverso, G., Shuber, A., Levin, B., et al. (2002) Detection of APC mutations in fecal DNA from patients with colorectal tumors. *N. Engl. J. Med.* **346,** 311–320.
8. Kobs, G., Hurst, R., Betz, N., et al. (2001) Fluorotect green lys in vitro translation labelling. *Promega Notes* **77,** 23–27.
9. Traverso, G., Diehl, F., Hurst, R. et al. (2003) Multicolor in vitro translation. *Nat. Biotechnol.* **21,** 1093–1097.
10. Rowan, A. J. and Bodmer, W. F. (1997) Introduction of a myc reporter tag to improve the quality of mutation detection using the protein truncation test. *Hum. Mutat.* **9,** 172–176.
11. Kahmann, S., Herter, P., Kuhnen, C., et al. (2002) A non-radioactive protein truncation test for the sensitive detection of all stop and frameshift mutations. *Hum. Mutat.* **19,** 165–172.
12. Kutzner N., Hoffmann I., Linke C., et al. (2005) Non-invasive colorectal cancer diagnosis by the combined application of molecular diagnosis and the faecal occult blood test. *Cancer Lett.* **229,** 33-41.
13. Kahmann, S. and Müller, O. A new method for detecting DNA mutations. US patent 101 07 317, pending.
14. Gite, S., Lim, M., Carlson, R., et al. (2003) A high-throughput nonisotopic protein truncation test. *Nat. Biotechnol.* **21,** 194–197.

# 9

# Creation of Novel Enantioselective Lipases by SIMPLEX

## Yuichi Koga, Tsuneo Yamane, and Hideo Nakano

## Summary

The single-molecule PCR-linked in vitro expression (SIMPLEX) technology, which can directly link a single molecule of a gene to its encoding protein, has been used to engineer enantioselectivity of lipase from *Burkhorderia cepacia* KWI-56. A combinatorial mutation has been introduced only to four residues in the hydrophobic substrate-binding pocket of the enzyme based on a structural model of the substrate–enzyme complex. Such focused mutation library constructed by the SIMPLEX technology has been screened for an enantiomeric substrate. Some combinations of substitutions in the four positions of the lipase have been found as effective for changing the enantio-preference from the ($S$)-form of $p$-nitrophenyl-3-phenylbutyrate to the ($R$)-form. Here, we describe the detail procedure to construct such an exclusively in vitro protein library and a practical screening method based on enzymatic activity.

**Key Words:** In vitro expression; HTS; lipase; enantioselectivity; molecular evolution; combinatorial mutant; SIMPLEX; single-molecule PCR.

From: *Methods in Molecular Biology*, vol. 375,
*In Vitro Transcription and Translation Protocols, Second Edition*
Edited by: G. Grandi © Humana Press Inc., Totowa, NJ

## 1. Introduction

Designing enantioselectivity of enzymes is one of the most attractive but challenging trials in the field of protein engineering for synthesis of enantiometically pure compounds, of which importance is expanding in pharmaceutical, agricultural, or synthetic organic chemistry. There is, however, no practical theory for introducing mutations into any enzyme to change their enantioselectivity. In recent works, a desired enantioselectivity has been given to enzymes by directed evolutional strategy, which comprises iterative cycles of mutation and identification of improved variants by screening or selection (1). For example, the enantioselectivity of hydantoinase has been inverted toward D,L-5-(2-methylthioethyl)hydantoin by error-prone PCR and following saturation mutagenesis (2), and enantioselectivity of a lipase from *Pseudomonas aeruginosa* was inverted by the combination of error-prone PCR and DNA shuffling (3).

These evolutional methods seem to be quite useful to improve some properties of proteins that cannot be easily obtained by rational approach. However, there are a number of limitations: use of living cells as hosts limits the variety of targeted proteins or screening methods; the library size is theoretically restricted by transformation efficiency; toxic proteins cannot be targeted; special in vivo assay methods are sometimes required; cell growth time could be a rate limiting. The single-molecule PCR-linked in vitro expression (SIMPLEX) system, of which theoretical issues are described in Chapter 4, can potentially overcome these limitations, because it allows the rapid preparation of protein libraries in vitro without any theoretical limitations (4). Almost all proteins may be targeted and almost all assay methods may be applied.

We have demonstrated the feasibility of the SIMPLEX technology, by showing the complete inversion of the enantioselectivity of the lipase from *Burkholderia cepacia* KWI-56 (lipase KWI-56) (5), which has high enantioselectivity for a model chiral compound, *p*-nitrophenyl-3-phenylbutyrate. This lipase is hardly expressed in *Escherichia coli* cells as active form, whereas it can be directly expressed from its gene with a modified in vitro expression system (6). Therefore, changing the enanitoselectivity of the lipase would be a good demonstration of the usefulness of SIMPLEX.

Mutation sites of the lipase has been selected based on a model for a reaction intermediate complex of the enantio-substrate and the lipase. Then a combinatorial mutation library of the lipase potentially including 2401 variants has been constructed by two-step SIMPLEX that comprises the first screening of a mutant library amplified from multiple template DNA molecules and the subsequent second screening of a library amplified from single DNA templates, resulting in the quick finding of variants with completely inverted enantioselectivity. Here, we will describe the practical details of the following procedures, (1) a mutation design based on a three-dimensional (3D) structural model of the substrate–lipase complex (2) the construction of the first and the second SIMPLEX libraries with combinatorial mutation and screening.

## 2. Materials

### *2.1. DNA Library (see Note 1)*

1. Plasmid: pRSET-rLip: a pRSET (Invitrogen, San Diego, CA) derivative containing mature lipase KWI56 gene between T7 promoter and T7 terminator *(6)*. It is used as template for mutation library.
2. Oligo DNA primers: K4 primer (TCA GAC TGG ACA CTA AAT GG) used as primer for Single Molecule PCR. This sequence was generated by "Homoprimer" software (run on MacOS 9 and available from the corresponding author) to avoid nonspecific binding or dimmer formation of primers *(7)*. F3-K4 (TCA GAC TGG ACA CTA AAT GGA CCG TAT TAC CGC CTT TGA GTG AG) and R3-K4 (TCA GAC TGG ACA CTA AAT GGA TCC GGA TAT AGT TCC TCC TTT CAG). These two primers are used for amplification of DNA library fragments containing, T7 promoter, wild-type or mutated mature lipase gene and T7 terminator, and also used for attaching the K4 sequence to both ends of the fragments. Primers for lipase gene mutagenesis: lipFL17VAG (CGA TCA TCC TCG TGC ACG GGG BCT CAG GTA CCG ACA AGT ACG C), lipFL17FILM (CGA TCA TCC TCG TGC ACG GGW TKT CAG GTA CCG ACA AGT ACG C-3'), lipFF119VAG (CGC CGC ATC GCG GCT CCG AGG BCG CAG ACT TCG TGC AGA ACG TGC), lipFF119FIL (CGC CGC ATC GCG GCT CCG AGW TKG CAG ACT TCG TGC AGA ACG TGC), lipFL167VAG (CGC TCG CCG CGC TGC AGA CGG BCA CGA CCG CCC GGG CTG CCA CG), lipFL167FIL (CGC TCG CCG CGC TGC AGA CGW TKA CGA CCG CCC GGG CTG CCA CG),

lipFL266VAG (GCT CCG GGC AGA ACG ACG GGG BCG TTT CGA AGT GCA GTG CGC TG), lipFL266FILM (GCT CCG GGC AGA ACG ACG GGW TKG TTT CGA AGT GCA GTG CGC TG), lipRL17 (CCC GTG CAC GAG GAT GAT CGG ATA ACG), lipRF119 (CTC GGA GCC GCG ATG CGG CGT GCC-3'), lipRL167 (CGT CTG CAG CGC GGC GAG CGC GTC C), lipRL266 (CCC GTC GTT CTG CCC GGA GCC GCG).

3. LA *Taq* DNA polymerase (Takara Shuzo, Kyoto, Japan).
4. TE buffer: 10 m$M$ Tris-HCl, 1 m$M$ EDTA, pH 7.5.
5. 0.01% Bluedextran 2000 (Amersham Pharmacia Biotech, Piscataway, NJ).
6. Ultra pure water.
7. GenElute™ agarose gel spin column (Sigma, St. Louis, MO).
8. Round-bottom 96-well plastic plate.
9. 384-Well PCR reaction plate.
10. GeneAmp PCR system 9700 (Applied Biosystems, Foster City, CA), or equivalent.
11. Microplate dispensing machine HT Station 500 (Cosmotec, Tokyo, Japan), or equivalent.

## 2.2. In Vitro Protein Expression (see Note 2)

1. 5 $M$ KOAc.
2. 0.6 $M$ Mg(OAc)$_2$.
3. 0.1 mg/mL Rifampicin (Sigma).
4. 10 mg/mL Creatine kinase (Roche Diagnostics).
5. T7 RNA polymerase (Takara Shuzo).
6. Triton X-100.
7. 3 mg/mL Activator protein: chaperon-like folding assisting protein specific for lipase KWI-56 (*see* **Note 3**).
8. *E. coli* S30 extract (*see* **Note 4**).
9. LM solution. For 390 μL of the LM solution, mix 40 μL of 2.2 $M$ Tris-acetate, pH 7.4, 9.5 μL of 200 m$M$ ATP, pH 7.4, 13.3 μL of 100 m$M$ each CGU mix, pH 7.4, 62.4 μL of 1.25 $M$ creatine phosphate Na$_2$, 15.6 μL of 20 amino acid mix (50 m$M$ each), 143 μL of 43.6% polyethylene glycol 8000 (Sigma), 5.4 μL of folinic acid calcium salt (Sigma), 15 μL of 17.4 mg/mL of *E. coli* tRNA (Roche Diagnostics), 40 μL of 1.4 $M$ ammonium acetate, and sterile water to 390 μL.
10. Flat-bottom 384-clear plate.

## 2.3. Lipase Activity-Based Screening

1. (*R*)- and (*S*)-*p*-Nitrophenyl-3-phenyl butyrate kindly supplied by Dr. Katsuya Kato (National Institute of Advanced Industrial Science and Technology, Japan).
2. Dimethyl sulfoxide.
3. 50 m*M* Potassium phosphate buffer pH 7.0 containing 1% Triton X-100.

# 3. Methods

## 3.1. Construction of Mutant Lipase Gene Library

### 3.1.1. Construction of a Structural Model of Reaction Intermediate

To design the mutation site, precise 3D structure is necessary. Unfortunately, the 3D structure of lipase KWI-56 is unknown, but it can be predicable by homology modeling. Because the amino acid sequence of lipase KWI-56 has 97% identity with that of the lipase from *B. cepacia*, whose structural data in an open conformation is available on PDB (2LIP), a structural model of lipase KWI-56 has been constructed based on the PDB data of *B. cepacia* lipase as follows:

1. Change all the different amino acid residues of *B. cepacia* lipase from lipase KWI-56 virtually on Insight II (Accelrys, San Diego, CA) to the corresponding amino acid residues of lipase KWI-56.
2. Do energy minimization of changed residues using the Discover software (Accelrys).
3. Make a covalent bond between the side chain oxygen of Ser 87 and the central carbon of the tetrahederal intermediate of (*S*)-configuration of 3-phenylbutyrate, then do the energy minimization and molecular dynamic simulation.

### 3.1.2. Design of Mutations

The strategy and design of mutation would be dependent on a desired property. In our case, because substrate specificity is the target, it seems very reasonable to focus on the substrate binding pocket. The geometric arrangement of the acyl-group of the substrate and the side chains of amino acid residues of the hydrophobic substrate binding pocket in the

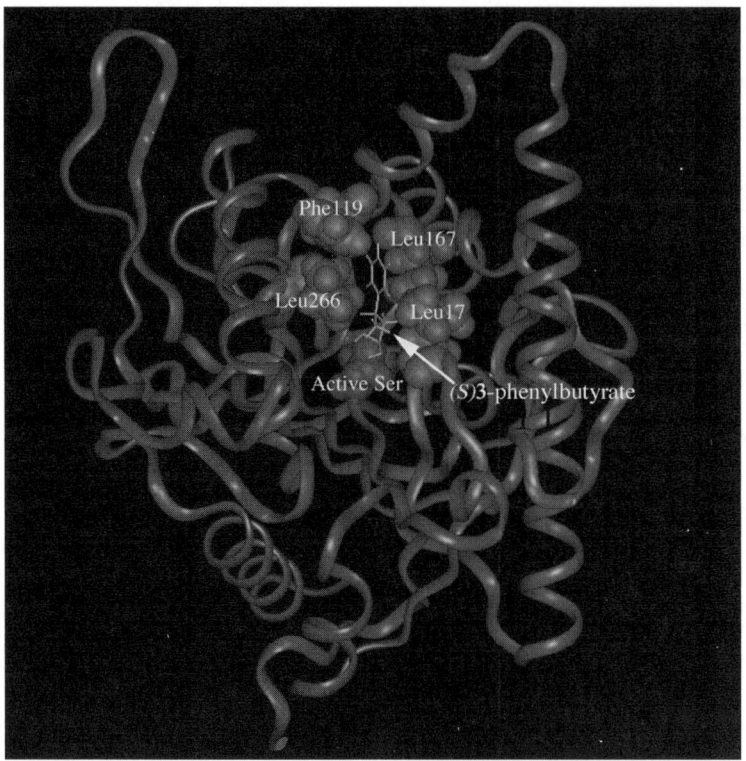

Fig. 1. A structural model of a reaction intermediate of lipase KWI-56. The structural model of the lipase was constructed with (*S*)-configuration 3-phenylbutyrate. The hydrophobic chain of the substrate is bound to the hydrophobic pocket of the lipase, and the bulky phenyl group is in proximity of Leu17, Phe119, Leu167, and Leu266.

structural model thus constructed show that four amino acid residues, L17, F119, L167, and L266, in the pocket are close to the acyl-group of the substrate (*see* **Fig. 1**). Therefore, these four amino acid residues have been selected as mutation sites. To keep the hydrophobicity of the pocket, the four residues have been substituted only for hydrophobic amino acid residues (G, A, V, L, I, M, F) in a combinatorial manner. Two degenerate codons, GBC and WTK were designed for Gly, Ala and Val for Phe, Leu, Ile and Met, respectively (B:T, C or G; W:T or A; K:T, or G ). Two mutation primers that include either of those two codons, and one reverse primer are used for each mutation site, so that totally 12 primers are used

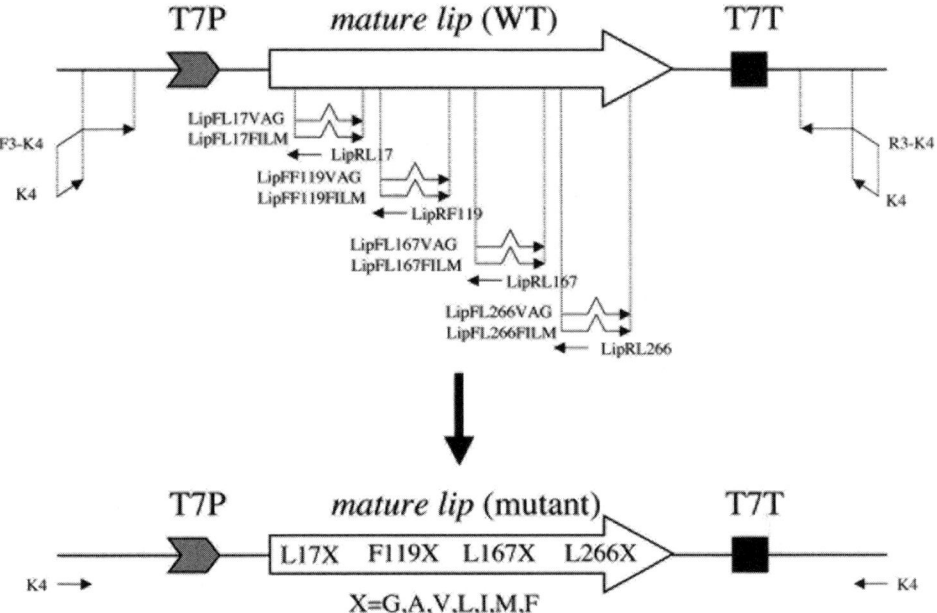

Fig. 2. Primers used for construction of a mutant lipase library. Mutations were introduced into four sites of lipase gene by PCR using primers with degenerate codons. K4 homo priming sequence was also introduced to both ends by F3-K4, and R3-K4 primers. The lipase gene fragments including various mutations were ligated to full length mature lipase gene by overlapping PCR and then reamplified by PCR using K4 homo primer. The full length PCR products include T7 promoter sequence and T7 terminator sequence at upstream and downstream of mutant lipase gene, respectively, so that those fragments can be used for in vitro transcription by T7 RNA polymerase.

for all the four mutation sites as illustrated in **Fig. 2**. The potential combination of mutation is up to 2401.

### 3.1.3. Construction of Mutant Lipase Genes

1. Five different regions of lipase KWI-56 gene that has overlapping sequences with neighbor regions were amplified with a pair of primers, one of which has semirandom mutations as shown in **Fig. 2**. Because the GC content of the lipase gene is high, the LA *Taq* polymerase were used under the conditions provided by the manufacturer *(8)*. pRSETrLip of 100 ng and 50 pmol each of primers were used for 100-µL scale PCR reaction.

2. Load each PCR products on 1% agarose gel for electrophoresis, cut out the DNA fragments of correct size from the gel, and purify it using GenElute™ Agarose gel spin column. The resultant solution is <20 µL for each fragments (*see* **Note 5**).

3. Mix equal amounts of all fragments (on a molar basis) together and carry out overlapping PCR using K4 primer to avoid amplification of template DNA.

4. Apply the resultant products on 1% agarose gel followed by electrophoresis, cut out and purify the DNA band of the predicted size of the lipase gene by GenElute Agarose gel spin column.

5. Concentrate the resultant DNA solution by ethanol precipitation with salts, and dissolve the DNA in 100 µL of TE buffer.

6. Determine the concentration of DNA by measuring the absorbance at 260 nm. For a lipase gene fragment of $8.5 \times 10^5$ Da, its concentration at $OD_{260}=1$ is $3.5 \times 10^{10}$ molecules/µL.

7. Dilute the DNA solution to the concentration of $10^{10}$ molecules/µL with TE buffer and distribute to new tube as 10 µL/tube and store at –20°C (*see* **Note 6**).

## *3.2. Construction of SIMPLEX Library for First Screening*

Although the library size of SIMPLEX is theoretically unlimited, in practical terms it is limited by the throughput of PCR thermocyclers. Therefore, a two-step SIMPLEX strategy has been developed *(8)*. (1) A first DNA library is produced by PCR from multimolecule and in vitro expression and screening, and (2) a second DNA library is produced from DNA molecules in positive wells by single molecule PCR, followed by the in vitro expression and screening. This two-step method can expand the searchable size of SIMPLEX library dramatically (*see* **Note 7**). In this trial, five molecules of the mutated lipase gene were amplified in 384-well plates and expressed in vitro in the first screening, resulting in quick screening of about 10,000 candidates. This number is large enough to cover all the possible combinations of mutations in this library.

### *3.2.1. "Five Molecule" PCR (see* **Note 8***)*

1. Prepare two 50-mL tubes filled with 40 mL of TE buffer containing 0.01% Bluedextraneach. Measure the weight of the tubes to know exactly the amount of each buffer. The weight of each tube should be measured previ-

ously. Bluedextran will prevent nonspecific binding of DNA to the wall of the tubes.

2. Add 4 µL of the DNA solution ($10^{10}$ molecules/µL) to one of the tubes containing 40 mL buffer and vortex. Estimate the exact dilution rate from the weight of the buffer. The concentration should be around $10^6$ molecules/µL.

3. Centrifuge the tube briefly to spin down all the liquid on the cap and wall of the tube to avoid dispersion of the DNA sample into the air.

4. Take out 4 µL of the diluted DNA solution and add them to another 40 mL buffer, and then vortex and spin down. The DNA concentration should be around 100 molecules/µL.

5. Prepare 6100 µL of PCR reaction mixture for two 384-well plates by mixing 610 µL of 10 × LA *Taq* buffer, 610 µL of 25 m*M* MgCl$_2$, 488 µL of 2.5 m*M* each dNTP mixture, 31 µL of 100 pmol/µL K4 primer, 44 µL of library the diluted DNA (100 molecules/µL)and 4229 µL of distilled water, keeping it on ice. Add 88 µL of LA *Taq* DNA polymerase just before dispensing. Because the amount of PCR reaction mixture for one well is 7 µL, the average amount of DNA molecule in 7 µL mixture should be five. Prepare 100 µL of PCR reaction mixtures both without DNA template and with wild-type lipase gene amplified from pRSETrLip by PCR with F3-K4 and R3-K4, as negative control and positive control of PCR, respectively.

6. Put 63 µL of the PCR mixture with the wild-type lipase gene into A1 well (left up corner) of 96-well plate, and put PCR mixture without DNA fragment into A2 well. These solutions were used as negative control of subsequent experiments. Then dispense 63 µL of PCR reaction mixture with diluted mutant DNA fragment to the rest of the wells in the plate using multichannel pipet and filter-tips. Keep the plate on ice while dispensing.

7. Dispense 7 µL of the reaction mixture from the 96-well plate to two 384-well plates using microplate dispensing machine HT Station 500. Seal the reaction plates after centrifuging briefly to push each sample to the bottom of each well.

8. Put the plates on GeneAmp PCR system 9700 and run the following temperature sequence: 5 s at 94°C; 65 cycles of 3 s at 98°C, 10 s at 50°C, and 80 s at 72°C; 7 min at 72°C, then store the plates at –20°C until use.

9. Repeat these procedures several times to get appropriate number of plates.

### 3.2.2. In Vitro Expression of Lipase

The DNA library is replicated by multidispenser and expressed in vitro from the plate. The in vitro expression mixture (4.2 µL) is mixed well with 1 μL of PCR product, totally 4.7 mL mixture for 2304-well will be prepared for screening.

1. Mix gently 350 μL of 5 *M* KOAc, 350 μL of Mg(OAc)$_2$, 1750 μL of 0.1 mg/mL rifampicin, 280 μL of 10 mg/mL creatin kinase, 50 μg of T7 RNA polimerase, 4375 μL of LM solution, 1750 μL of 3 mg/mL activtor protein, 175 μL of Triton X-100, and 4987.5 μL of S30 *E. coli* cell extract in the 50-mL tube by inverting the tube. These reagents should be mixed just before use.
2. The in vitro expression mixture is dispensed to 151 μL/well on a 96-well plate by multichannel pipet, and then dispensed again to six new 384-clear plates using dispensing machine HT Station 500–4.2 μL/well (**Fig. 3**) (*see* **Note 9**).
3. Transfer 1 μL/well of PCR product from the PCR plates to the plate with the in vitro expression reaction premixture in **step 2**.
4. Seal the plate and incubate at 30°C for 2.5 h.
5. The expressed lipase variants should be stored at 4°C and used for activity assay immediately.

## 3.2.3. Screening Based on Lipase Activity Against (R)-p-Nitrophenyl-3-Phenylbutyrate

The mutant lipase variants are screened for (*R*)-*p*-nitrophenyl-3-phenylbutyrate, which is hardly hydrolyzed by the wild-type lipase, so that the yellow color of *p*-nitrophenol is observed only when the lipase variants with changed enantioselectivity exist in the well.

1. (*R*)-*p*-nitrophenyl-3-phenylbutyrate (25.7 mg) is dissolved completely with 9 mL of DMSO, and then the solution is diluted with 81 mL of 50 m*M* potassium phosphate buffer (pH 7.0) containing 1% of Triton X-100. The concentration of the substrate would be 1 m*M*. This is the volume for 3000 assays.
2. Dispense the substrate solutions to three 96-well plates (300 μL/well, totally 86.4 mL).
3. Dispense 30 μL of the substrate solution to each well of the 384 plates expressing the lipase variants using an automatic dispenser (**Fig. 3**).
4. Start incubation of the 384 plates at 37°C immediately after the addition of the substrate.
5. After 1-h incubation, measure absorbance of each reaction mixture at 405 nm by a plate reader.

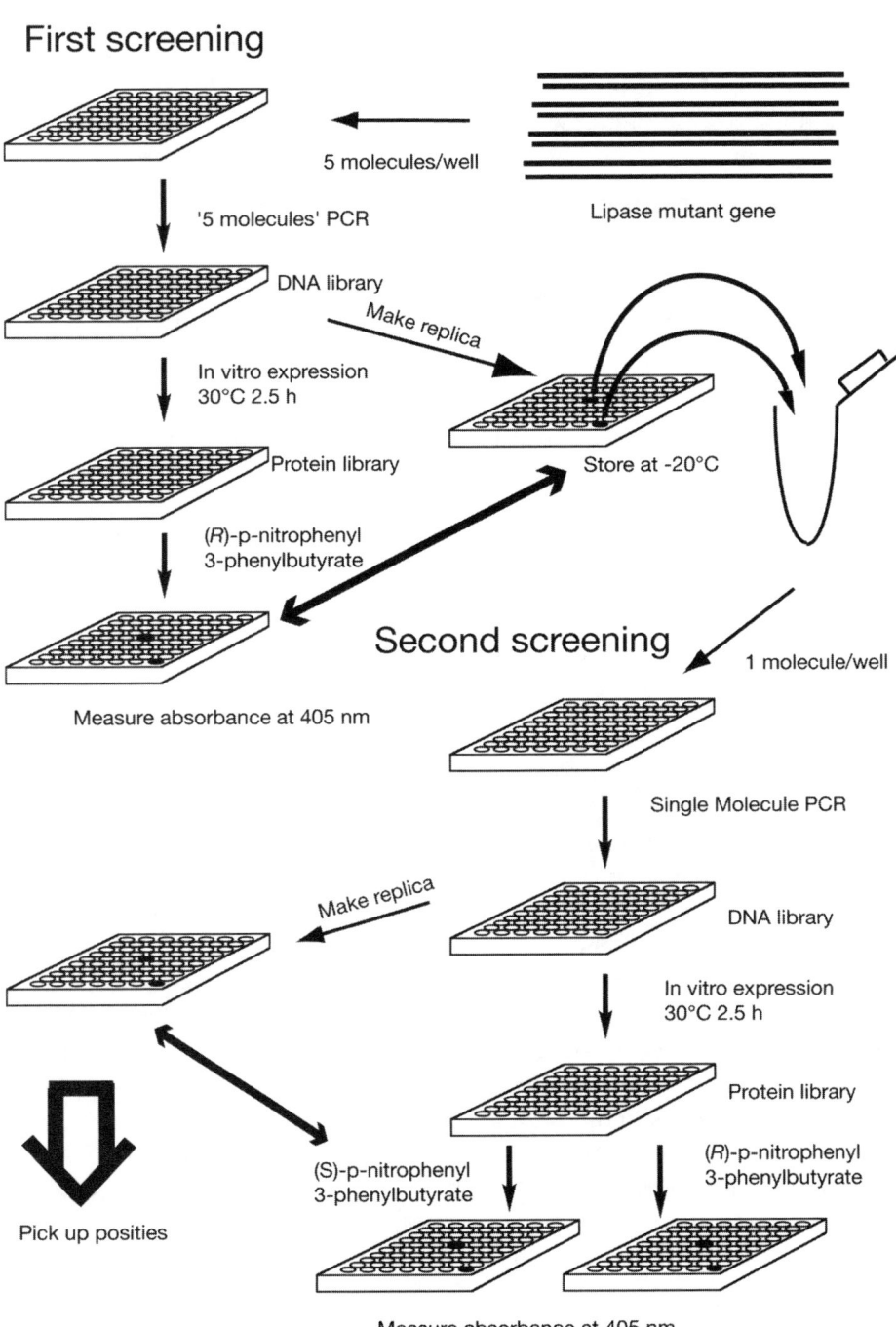

Fig. 3. A schematic drawing of library construction and screening.

6. Identify the positive wells (plate number and location of well in the plate) showing higher absorbance than the WT control, and recover the DNA from the corresponding well of DNA library replica plates.

## 3.3. Second Screening of Library Made by SIMPLEX

### 3.3.1. Construction of SIMPLEX Library (see **Note 8**)

The lipase variants that can hydrolyze (*R*)-*p*-nitrophenyl-3-phenylbutyrate should be included in the pool of the first screening. The next step is to express these candidates separately by SIMPLEX and to measure their activity for (*R*) and (*S*)-*p*-nitrophenyl-3-phenylbutyrate are measured.

1. Mix the DNA solutions recovered from the positive wells in the first screening and precipitate with ethanol and then dissolved in 100 µL of TE buffer.
2. Determine the concentration of the DNA mixture from absorbance of 260 nm as in **Subheading 3.1.3.**, **step 6**. Adjust the concentration of the DNA solution to $1 \times 10^{10}$ molecules/µL with TE buffer, and then stored at –20°C.
3. Dilute the $1 \times 10^{10}$ molecules/µL DNA solution with TE buffer (pH 7.5) containing 0.01% bluedextran as described in **Subheading 3.2.1.**, **steps 1–4** to 100 molecules/µL.
4. Twenty microliters of the DNA solution was diluted with 80 µL TE buffer containing blue dextran to make the 20 molecules/µL solution.
5. Prepare 6100 µL of PCR reaction mixture in a new tube as follows. Six hundred-ten microliters of 10X LA *Taq* buffer, 610 µL of 25 m*M* MgCl$_2$, 488 µL of 2.5 m*M* each dNTP mixture, 31 µL of 100 pmol/µL K4 primer, 44 µL of 100 molecules/µL library DNA and 4229 µL distilled water. Mix well and keep on ice. Add 88 µL of LA *Taq* DNA polymerase just before dispensing. The concentration of DNA fragment should be adjusted so as to be 1 molecules in one well according to the volume of each reaction.
6. Dispense 7 µL of the PCR mixture to each well of two 384-well plates as described in **Subheading 3.2.1.** The average number of molecules in each well should be one.
7. Amplify the DNA fragments in the plates with the following thermal sequence: 5 s at 94°C, 65 cycles of 3 s at 98°C, 10 s at 50°C, 80 s at 72°C, and 7 min at 72°C. Analyze some of the PCR reactions by agarose gel electrophoresis to check amplification efficiency (*see* **Note 7**).

8. Make replica plates by transferring 1 μL of each PCR products to a new 384 plates using HT station 500. The replica plates should be used for further expression assay and the original master plates should be stored at –20°C.

9. Prepare the in vitro expression reagents mixture for 1536 samples as follows: mix 220 μL of 5 *M* KOAc, 220 μL of 0.6 *M* Mg(OAc)₂, 1100 μL of 0.1 mg/mL rifampicin, 176 μL of 10 mg/mL creatin kinase, 55 μL of 2.3 mg/mL T7 RNA polymerase, 2750 μL of LM solution, 1100 μL of 3 mg/mL activator protein, 110 μL of Triton X-100, and 3135 μL of S30 *E. coli* cell extract. Its total volume (8866 μL) includes the dead volume of dispensing machine as well as sample mixture.

10. Add 4.2 μL of the in vitro expression reagents to each well of the replica DNA library plates, and incubate at 30°C for 2.5 h (**Fig. 3**).

## 3.3.2. Screening of Secondary SIMPLEX Library

The secondary SIMPLEX library on replica plates should be screened for both (*R*) and (*S*)-*p*-nitrophenyl-3-phenylbutyrate hydrolysis activity.

1. Prepare 30 mL of substrate solutions containing either 1 m*M* (*R*) or (*S*)-*p*-nitrophenyl-3-phenylbutyrate for 768 samples as follows: 8.6 mg of (*R*) or (*S*)-*p*-nitrophenyl-3-phenylbutyrate should be dissolved in 3 mL of DMSO completely, and dilute with 27 mL of 50 m*M* potassium phosphate, pH 7.0, containing 1% Triton X-100.

2. Add 30 μL of the (*R*)-form substrate solution to each well on one replica plates and the same amount of the (*S*)-form substrate solution to each well of a second replica plate.

3. Plates should be incubated at 37°C for 2 h.

4. Measure the absorbance at 405 nm by a microplate reader FLUO star Galaxy (BMG, Germany).

## 3.3.3. Data Analysis and Following Up of the Screening

More than 20 variants hydrolyzed the (*R*)-substrate whereas the WT lipase had a strong preference for the (*S*)-form substrate. Furthermore, some variants had a higher activity for the (*R*)-form substrate than the (*S*)-form substrate, suggesting that the enantioselectivity of the lipase can be successfully inverted from the (*S*)-form substrate to the (*R*)-form one. The selected lipase variants in the second library were cloned and

**Table 1**
**The Amino Acid Substitutions and Relative Activity for (R)- and (S)-**
**Configuration of p-Nitrophenyl 3-Phenylbutyrate of Lipase Variants (5)**

| | | | | | Relative activity | |
|---|---|---|---|---|---|---|
| Variants no. | | Amino acid substitutions | | | *R* | *S* |
| WT | L17 | F119 | L167 | L266 | 0.03 | 1 |
| 1 | F | L | G | V | 1.35 | 0.22 |
| 2 | F | L | A | V | 1.23 | 0.3 |
| 3 | F | M | G | A | 1.15 | 0.19 |
| 4 | F | V | G | V | 1.15 | 0.56 |
| 5 | A | F | G | I | 1.06 | 0.52 |
| 6 | L | L | G | V | 0.76 | 0.34 |
| 7 | F | M | A | V | 0.73 | 0.37 |
| 8 | A | A | V | V | 0.58 | 0.14 |
| 9 | G | F | A | L | 0.48 | 0.83 |
| 10 | G | L | A | I | 0.42 | 0.47 |
| 11 | M | M | G | V | 0.4 | 0.15 |
| 12 | M | L | A | V | 0.34 | 0.36 |
| 13 | F | F | A | L | 0.32 | 0.12 |
| 14 | A | F | A | V | 0.19 | 0.42 |
| 15 | M | A | G | V | 0.14 | 0.19 |
| 16 | A | L | G | A | 0.08 | 0.04 |
| 17 | G | F | G | V | 0.05 | 0.3 |

sequenced (**Table 1**). Moreover, four of the most active variants were expressed and purified for enzymatic analysis. From the biochemical analysis using a chiral substrate of *p*-nitrophenyl-3-phenylbutyrate and high-performance liquid chromatography (HPLC) analysis, those lipase variants had inverted enantioselectivity (**Table 2**) (*6*).

## 3.4. Concluding Comments

Because all the screened lipase variants showing reversed enantioselectivity had more than three mutations and each single mutation did not change its enantioselectivity, the combination of mutations may be required for the inversion. These combinations of several amino acid mutations are, however, hardly obtainable by the error-prone PCR because introducing more than one amino acid replacements in a specific

**Table 2**
**Summary of Purified Lipase Variants With Reversed Enantioselectivity** *(5)*

| Variant no. | Mutations | Specific activity (U/mg) | Enantioselectivity (conversion) |
|---|---|---|---|
| WT | – | 60.2 3.1 | ES=33 1 (47%) |
| 11 | L17F, F119L, L167G, L266V, (T251A)[a] | 90.6 2.9 | ER= 38 2 (47%) |
| 2 | L17F, F119L, L167A, L266V, (D21N)[a] | 16.6 0.6 | ER= 33 1 (49%) |

[a]Indicates mutations outside of the targeted site.

position is almost impossible. Therefore, a focused combinatorial mutation library has been proved to be effective to give a novel selectivity to enzymes.

## 4. Notes

1. All reagents should be carefully prepared to avoid nuclease and specific DNA contamination. Particularly, because the single-molecule PCR is very sensitive, special care should be taken to avoid contamination of DNA via tip, pipet, and hands. It is recommended to dispense a small amount of each reagent to separate tubes for storage before starting experiments.
2. *E. coli* S30 cell extract is commercially available from many suppliers such as Roche Diagnostics and Promega (Madison, WI).
3. The activator protein should be expressed in *E. coli* BL21(DE3) harboring pRSET-Act *(5)*, a T7 expression vector containing activator gene from *B. cepacia* KWI-56, and purified from the crude extract of the *E. coli* cells by precipitation with 20% saturated ammonium sulfate and following dialysis against 50 m$M$ Tris-HCl, pH 7.4, containing 10 m$M$ Mg(OAc)$_2$.
4. If you need in-house extract, please refer to **ref. 9**.
5. The purpose of purification of amplified fragments is to remove template gene and primers. It is important for efficient mutation.
6. Highly concentrated DNA solution might cause cross contamination in the amplification from single molecules. Store DNA solution separately, and do not share storage boxes with reagents used for the single-molecule PCR. In addition, avoid to use the same bench, pipet, tube folder, and reagents for template preparation and the single-molecule PCR.

7. When DNA fragments are dispensed on an average of one molecule per well, theoretically the DNA fragments are present in 63.2% of the wells, according to the Poisson distribution. It means that 141 wells of each 384-well plate are empty and cannot be used for screening.

8. Because the PCR procedure used in the following section is extremely sensitive, cross contamination of DNA to the reaction tube will easily give serious damage to your experiment. Filter tips are recommended to dilute and transfer DNA solutions.

9. Because the in vitro expression reagents contain some insoluble components, mix gently by inverting the tube to make homogeneity before dispensing to a 96-well plate.

## References

1. Robertson, D. E. and Steer, B. A. (2004) Recent progress in biocatalyst discovery and optimization. *Curr. Opin. Chem. Biol.* **8,** 141–149.

2. May, O., Nguyen, P. T., and Arnold, F. H. (2000) Inverting enantioselectivity by directed evolution of hydantoinase for improved production of L-methionine. *Nature Biotechnol.* **18,** 317–320.

3. Zha, D., Wilensek, S., Hermes, M., Jaeger, K. E., and Reetz, M. T. (2001) Complete reversal of enantioselectivity of an enzyme-catalyzed reaction by directed evolution. *Chem. Commun.* **24,** 2664–2665.

4. Rungpragayphan, S., Kawarasaki, Y., Imaeda, T., Kohda, K., Nakano, H., and Yamane, T. (2002) High-throughput, cloning-independent protein library construction by combining single-molecule DNA amplification with *in vitro* expression. *J. Mol. Biol.* **318,** 395–405.

5. Koga, Y., Kato, K., Nakano, H., and Yamane, T. (2003) Inverting enantioselectivity of *Burkholderia cepacia* KWI-56 lipase by combinatorial mutation and high-throughput screening using single-molecule PCR and *in vitro* expression. *J. Mol. Biol.* **331,** 585–592.

6. Yang, J., Kobayashi, K., Iwasaki, Y., Nakano, H., and Yamane, T. (2000) *In vitro* analysis of roles of disulfide bridge and calcium binding site in activation of *Pseudomonas* sp. strain KWI-56 lipase. *J. Bacteriol.* **182,** 295–302.

7. Nakano, H., Kobayashi, K., Ohuchi, S., Sekiguchi, S., and Yamane, T. (2000) Single-step single-molecule PCR of DNA with a homo-priming sequence using a single primer and hot-startable DNA polymerase. *J. Biosci. Bioeng.* **90,** 456–458.

8. Rungpragayphan, S., Nakano, H., and Yamane, T. (2003) PCR-linked in vitro expression: a novel system for high-throughput construction and screening of protein libraries. *FEBS Lett.* **540,** 147–150.
9. Ellman, J., Mandel D., Anthony-Cahill, S., Noren, C. J., and Schultz, P. G. (1991) Biosynthetic method for introducing unnatural amino acids site-specifically into proteins. *Methods Enzymol.* **202,** 301–337.

# 10

## In Vitro Transcription and Translation Coupled to Two-Dimensional Electrophoresis for Bacterial Proteome Analysis

Nathalie Norais, Ignazio Garaguso, Germano Ferrari, and Guido Grandi

### Summary

The most popular approach for proteomic analysis is based on the combination of two-dimensional electrophoresis (2DE) and mass spectrometry. Although very effective, the method suffers from a number of limitations, the most serious one being the necessity of expensive and sophisticated instrumentation to be operated by skilled personnel. Here, we propose an alternative approach, which is particularly useful when one is interested to establish if a subset of proteins is present in a complex protein mixture derived from a sequenced organism. The method is based on amplification of the genes whose products are under investigation. The amplified genes are used in transcription and translation reactions and the derived radio-labeled proteins are separated by 2DE. The gel is autoradiographed and the autoradiograph is superimposed on the 2D gel (sample gel) from which the protein mixture from the organism has been separated. The matching between the autoradiographic spots and the protein spots of the sample gel allows immediate protein identification.

**Key Words:** Bacterial proteomics; two-dimensional electrophoresis; in vitro transcription-translation; *Neisseria meningitidis* B.

From: *Methods in Molecular Biology,* vol. 375,
*In Vitro Transcription and Translation Protocols, Second Edition*
Edited by: G. Grandi © Humana Press Inc., Totowa, NJ

# 1. Introduction

Global analysis of proteins from cells and tissues is termed "proteomics." Although a variety of alternative procedures have been developed, at present one of the most popular approaches for proteomic analysis requires the previous knowledge of the genome sequence of the organism under investigation and is based on the combination of two-dimensional gel electrophoresis (2DE) and mass spectrometry. According to this approach, proteins are separated by 2DE, stained, in-gel digested with trypsin or other proteolytic enzymes and finally subjected to matrix-assisted laser desorption/ionization time-of-flight (MALDI-TOF) mass spectrometry. The MALDI-TOF analysis provides peptide mass fingerprints, which lead to protein identification when compared with the theoretical *in silico* fingerprints generated from the available genome sequence. Usually, 80–85% of the analyzed protein spots give a mass fingerprint, which is in most cases sufficient for protein identification. A limited number of spots (approx 5%) requires further tandem mass spectrometry analysis (MS/MS) for unambiguous characterization.

The major drawback of this approach is that it requires expensive and sophisticated instruments, which need to be operated by well-trained and specialized scientists. In addition, the method presents some limitations in sensitivity, not owing to mass spectrometers (which can analyze samples in the low fmole range), but to sample preparation procedures, which are usually inefficient, making the analysis feasible only when protein quantities greater than 0.1–0.2 pmole are available *(1)*. We have recently described an alternative method for the characterization of proteomes, in particular bacterial proteomes, which may offer some advantages over the current proteomic approaches *(2)*. The method combines PCR cloning, in vitro transcription-translation and 2DE. In this chapter, we describe two applications of the method, one aimed at identifying a single protein in the two-dimensional (2D) maps, the second designed to simultaneously identify a set of proteins.

## 1.1. Single Protein Identification in 2D Maps

When the scope of the investigation is to establish whether a specific protein is present in a complex protein mixture, for instance in the total protein extract of a given bacterium, the following procedure is proposed

(**Fig. 1**). The bacterium is grown under appropriate conditions and the bacterial culture is used to (1) prepare the mixture of cellular proteins (**Fig. 1**, step 1a) and (2) purify chromosomal DNA (**Fig. 1**, step 1b). Chromosomal DNA is used to amplify the gene encoding the protein under investigation (**Fig. 1**, step 2) and the amplified gene is used to drive the in vitro expression of the radio-labeled protein (**Fig. 1**, step 3). The transcription and translation reaction (TTR) is analyzed by sodium dodecyl sulfate-polyacrylamide gels (SDS-PAGE) for a qualitative and quantitative evaluation (**Fig. 1**, step 4). TTR is then mixed with the total cellular proteins and the mixture is separated by 2DE (**Fig. 1**, step 5). The 2D gel is then stained with Coomassie Blue (**Fig. 1**, step 5). Because a relatively small amount of TTR is added to the bacterial total protein sample (the proper amount of TTR is estimated by analytical monodimensional electrophoresis and autoradiography, and corresponds to the amount sufficient to visualize the labeled translated protein after overnight exposure) the proteins of the *Escherichia coli* S-30 cell extracts do not show up on the gel and the only visible spots derive from the bacterial protein mixture under examination. After staining, the 2D gel is autoradiographed (**Fig. 1**, step 6) and the autoradiograph is finally superimposed on the stained gel (**Fig. 1**, step 7). The protein spot, which eventually matches the spot visible on the autoragiograph, corresponds to the protein encoded by the amplified gene used in the TTR.

## 1.2. Multiple Protein Identification in 2D Maps

The procedure described for single protein identification is not amenable for the simultaneous identification of several protein spots. In fact, this would require the addition of several TTRs to the protein mixture under analysis. In so doing, the amounts of S-30 *E. coli* proteins would be high enough to be visualized after Coomassie Blue staining, thus making the subsequent identification of the protein spots in the sample mixture very complicated.

To overcome this problem, the following procedure has been developed (**Fig. 2**). A bacterial culture is used for protein sample (**Fig. 2**, step 1a) and chromosomal DNA preparation (**Fig. 2**, step 1b). Chromosomal DNA is used for gene amplification and the amplified genes are either cloned in appropriate expression vectors or utilized as linear fragments for in vitro

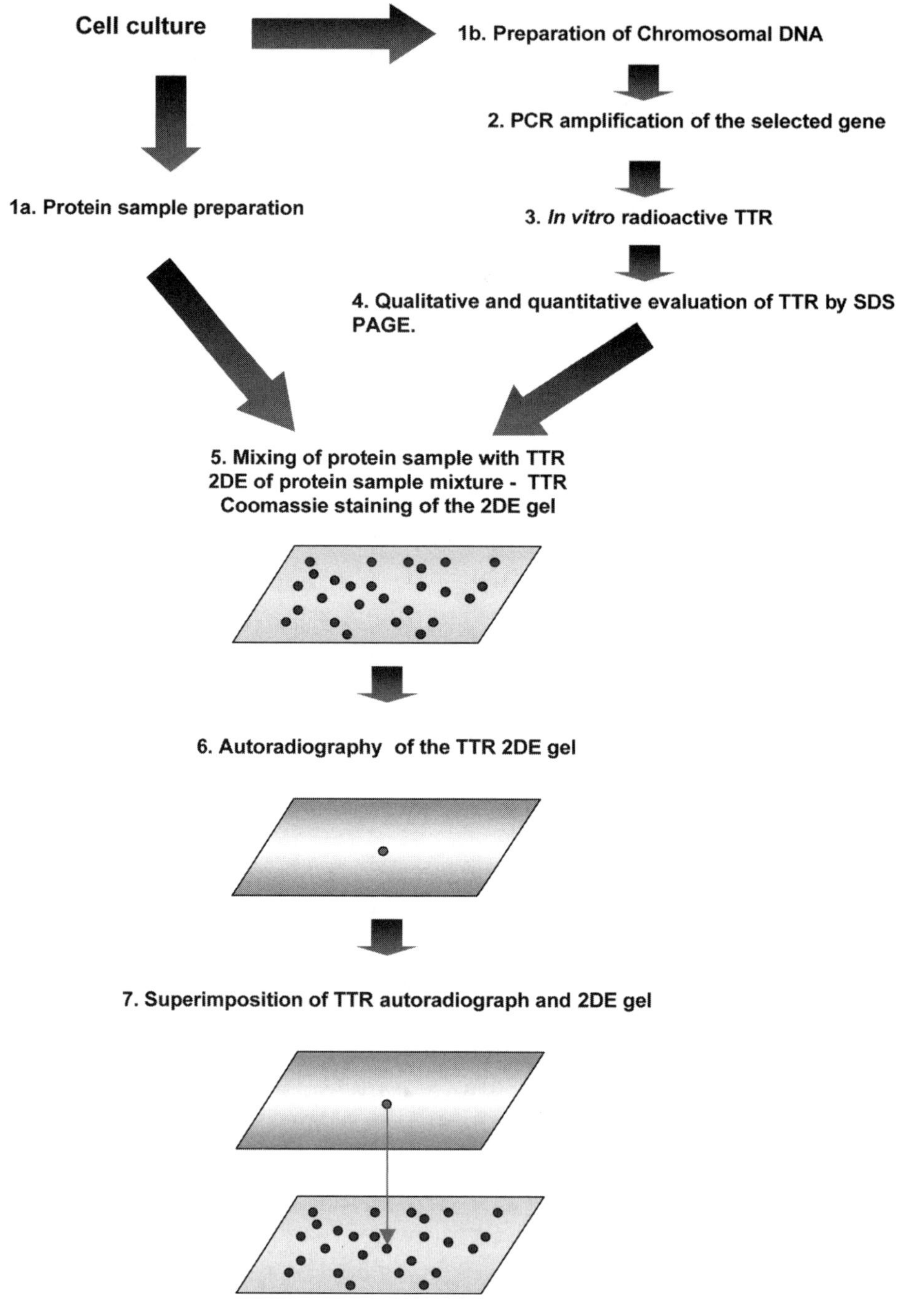

**Cell culture** ➡ **1b. Preparation of Chromosomal DNA**

**2. PCR amplification of the selected gene**

**1a. Protein sample preparation**                    **3. *In vitro* radioactive TTR**

**4. Qualitative and quantitative evaluation of TTR by SDS PAGE.**

**5. Mixing of protein sample with TTR**
**2DE of protein sample mixture - TTR**
**Coomassie staining of the 2DE gel**

**6. Autoradiography of the TTR 2DE gel**

**7. Superimposition of TTR autoradiograph and 2DE gel**

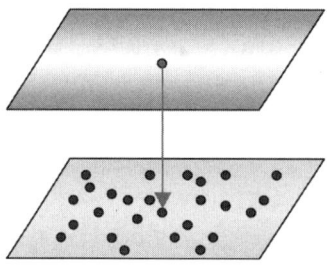

Fig. 1

protein synthesis (**Fig. 2**, step 2). In vitro TTRs are carried out in the presence of $^{14}$C-labeled amino acids (**Fig. 2**, step 3). Small aliquots of each reaction are analyzed by SDS-PAGE for a qualitative and quantitative evaluation, and TTRs are properly pooled on the basis of the SDS-PAGE analysis (**Fig. 2**, step 4). Pooled TTRs and the protein sample are separately resolved by 2DE in the presence of the same protein markers, and the 2D gels are Coomassie Blue-stained (**Fig. 2**, step 5). The 2DE gel of the TTRs is then dried and autoradiographed for TTR visualization, and the position of the protein markers are reported on the autoradiograph (**Fig. 2**, step 6). Finally, the TTR autoradiograph and sample gel are superimposed by computer-assisted image analysis using the protein markers as landmarks (**Fig. 2**, step 7). Spot matching between radioactive spots on the autoradiograph and Coomassie-stained protein spots on the protein sample allow the identification of protein spots on the sample gel (*see* **Note 1**).

In the following sections, we will first provide detailed protocols for multiprotein identification in 2D maps and then the methods for single protein identification.

---

Fig. 1. *(opposite page)* Schematic representation of the procedure for the identification a single specific protein 2D map. The protein mixture under investigation (protein sample, step 1a) and chromosomal DNA (step 1b) are prepared starting from the same bacterial culture. Chromosomal DNA is used for the amplification of the gene whose product is under investigation and the amplified gene is either cloned in an appropriate expression vector or utilized as linear fragment for in vitro protein synthesis (step 2). TTR is carried out in the presence of $^{14}$C-labeled leucine and lysine (step 3). A small aliquot of the reaction is analyzed by SDS-PAGE for a qualitative and quantitative evaluation (step 4). The protein sample obtained from the bacterial culture is then mixed with a small aliquot of TTR, and the mixure is resolved by two-dimensional electrophoresis (step 5). The 2D-gel is stained with Colloidal Coomassie Blue, and then dried and autoradiographed (step 6). Finally, autoradiograph and gel are superimposed (step 7). The gel spot that matches with the spot on the autoradiography corresponds to the protein encoded by the amplified gene used in the TTR.

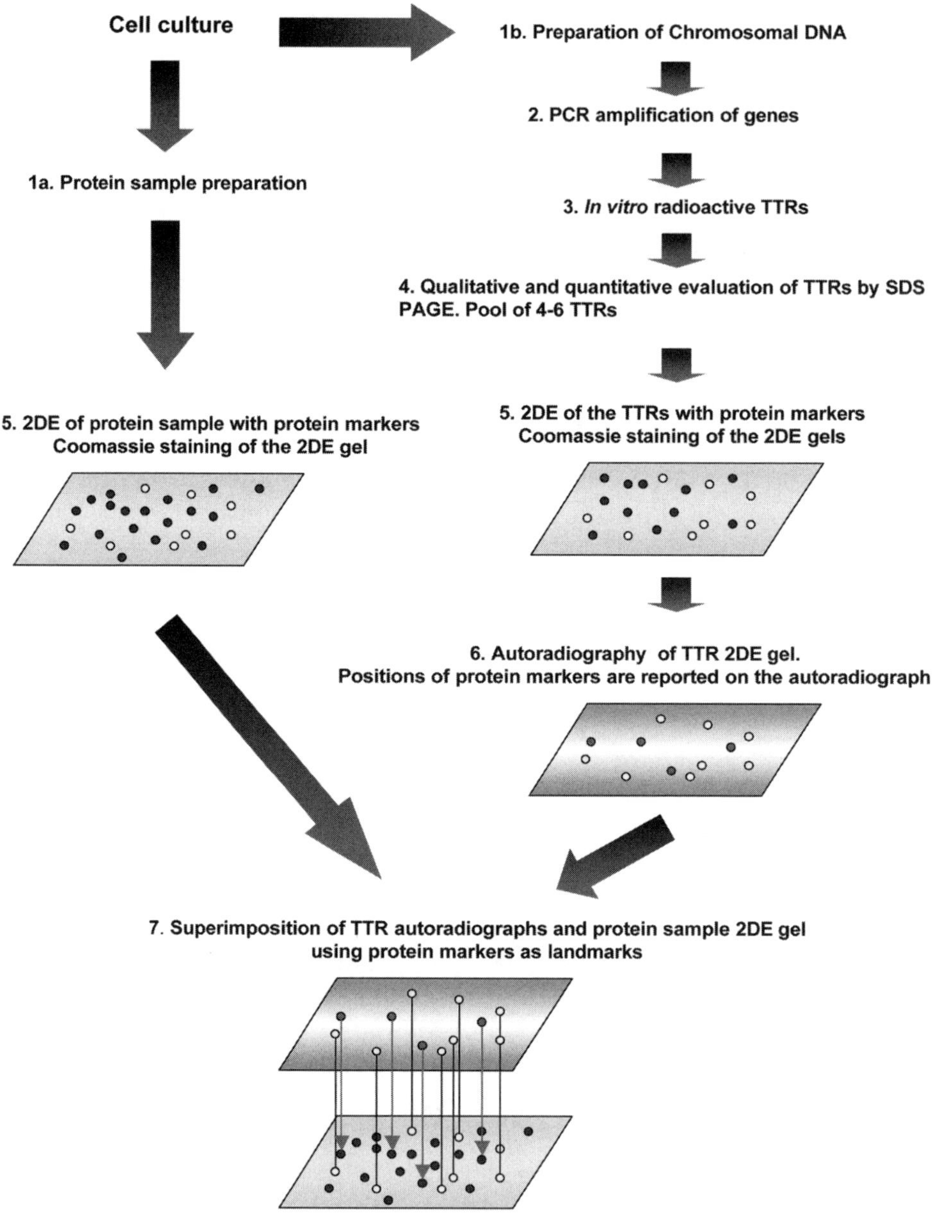

**Cell culture**        **1b. Preparation of Chromosomal DNA**

**2. PCR amplification of genes**

**1a. Protein sample preparation**

**3.** *In vitro* **radioactive TTRs**

**4. Qualitative and quantitative evaluation of TTRs by SDS PAGE. Pool of 4-6 TTRs**

**5. 2DE of protein sample with protein markers Coomassie staining of the 2DE gel**

**5. 2DE of the TTRs with protein markers Coomassie staining of the 2DE gels**

**6. Autoradiography of TTR 2DE gel. Positions of protein markers are reported on the autoradiograph**

**7. Superimposition of TTR autoradiographs and protein sample 2DE gel using protein markers as landmarks**

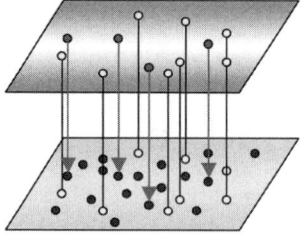

Fig. 2

## 2. Materials

### 2.1. Device, Growth Medium, and Buffers for Neisseria meningitidis Outer Membrane Protein

1. Device for bacteria lysis. Bacterial cells are disrupted with a French Press apparatus (SLM Instruments, Inc., Rochester, NY).
2. Bacteria growth medium. Bacteria are grown on GC agar plates (BD Biosciences, Franklin Lakes, NJ) supplemented with 4 g/L glucose, 0.1 g/L glutamine, and 2.2 mg/L cocarboxylase at 37°C in a humidified atmosphere containing 5% $CO_2$.
3. Bacteria lysis and wash buffers. Bacterial cells were washed in 40 m$M$ Tris-base and cell lysis was carried out in 40 m$M$ Tris-base containing 1000 U of Benzonase (Sigma, St. Louis, MO).

---

Fig. 2. *(opposite page)* Schematic representation of the strategy for multiple protein identification on a 2D map. The protein mixture under investigation (protein sample, step 1a) and chromosomal DNA (step 1b) are prepared starting from the same bacterial culture. Chromosomal DNA is used for the amplification of the genes of interest and the amplified genes are added to in vitro transcription and translation reactions for the synthesis of the corresponding proteins (step 2). TTRs are carried out in the presence of [14]C-labeled leucine and lysine (step 3). Small aliquots of each reaction are analyzed by SDS-PAGE for a qualitative and quantitative evaluation, and TTRs are properly pooled on the basis of the SDS-PAGE analysis (step 4). Pooled TTRs and protein sample are separately resolved by 2DE in the presence of protein markers, and the 2D gels are stained with Coomassie Blue (step 5, protein marker spots represented by open circles [yellow]). The 2D gel on which the mixture of the transcription and translation reactions (TTRs) has been resolved is dried and autoradiographed for the visualization of the radioactive in vitro synthesized proteins (gray circles [red], step 6). The position of the protein markers (open circles [yellow]) are reported on the autoradiograph (step 6). Finally, the TTR autoradiograph and sample gel are superimposed by computer-assisted image analysis using the protein markers as landmarks (step 7). The protein spots that match with the radioactive spots correspond to the products of the genes used for the TTRs.

## 2.2. PCR Amplification and Cloning of the Genes of Interest

PCR amplification is performed with Pwo DNA polymerase (Roche Diagnostics GmbH, Mannheim, Germany). Genes of interest were cloned into the plasmid pET-21b+ (Novagen, Madison, WI). Plasmid preparations were carried out using the Qiagen kit (Qiagen GmbH, Hilden, Germany).

## 2.3. 2DE Devices

First-dimension isoelectric focusing is performed using devices from Amersham Biosciences (Uppsala, Sweden). The devices include an Immobiline Dry-Strip Reswelling Tray, IPGphor strip holders, and an IPGphor Isoelectric Focusing System. Second-dimension Gradient polyacrylamide SDS-PAGE electrophoresis is performed using devices from Bio-Rad (Hercules, CA). The devices include a Mini-Protean II Multi-Casting Chamber, Model 485 Gradient Former and a Mini-Protean II Cell.

## 2.4. Mixture for In Vitro Transcription/Translation Reactions

1. $^{14}$C-labeled amino acids. L-[U-$^{14}$C] leucine and L-[U-$^{14}$C] lysine at 11.7 and 12.2 Gbq/mmol, respectively, are purchased from Amersham Biosciences.
2. Amino acid mixture. The mixture includes each of the 20 amino acids, with the exception of leucine and lysine, at a concentration of 1 m$M$ in diethylpyrocarbonate (DEPC)-treated water. The amino acid mixture is stable for more than 1 yr when stored at –20°C. DEPC-treated water is prepared by adding 1 mL of DEPC per liter of bidistilled water; the solution is stirred for 1 h and then autoclaved.
3. Low molecular weight (LMW) mixture. A 470-µL stock of LMW mixture can be prepared and stored in 100-µL aliquots at –20°C. It will be stable for several months, and can be thawed and frozen a few times. The composition of the LMW is prepared from stock solutions as described in **Table 1**. All stock solutions are kept at –20°C and are stable for several months. The stock solutions labeled with an asterisk are autoclaved.
4. Reaction mixture. The reaction mixture is prepared before use as described in **Table 2**.

**Table 1**
**LMW Mixture**

| Stock solution | Volume (μL) |
| --- | --- |
| 2.2 *M* Tris-acetate pH 8.2[a] | 40 |
| 0.55 *M* DTT | 5 |
| 38 m*M* ATP pH 7.0 | 50 |
| 88 m*M* CTP pH 7.0 | 15 |
| 88 m*M* GTP pH 7.0 | 15 |
| 88 m*M* UTP pH 7.0 | 15 |
| 0.42 *M* Phosphoenol pyruvate pH 7.0 | 100 |
| 40% Polyethylene glycol-6000[a] | 75 |
| 2.7 mg/mL Folinic acid | 20 |
| 17.4 mg/mL Transfer (t)RNA | 15 |
| 1.4 *M* Ammonium acetate[a] | 40 |
| 2.8 *M* Potassium acetate[a] | 40 |
| 0.38 *M* Calcium acetate[a] | 40 |

[a]Autoclaved stock solutions.

**Table 2**
**Reaction Mixture**

| | Volume (μL) for 10 TTRs |
| --- | --- |
| Amino acid mixture (*see* **Subheading 2.4.**) | 32.6 |
| 280 m*M* Mg acetate | 2.8 |
| 10 mg/mL Rifampicine in methanol | 2.2 |
| LMW mixture (*see* **Subheading 2.4.**) | 16.3 |
| S30-extract | 26.1 |

## 2.5. Mixtures for 2DE

1. Acrylamide mixture. The acrylamide mixtures are prepared before use; the amounts given are necessary for the preparation of 12 gels (1.5 mm thick, 7.3 cm high, 8.3 cm wide). Fifty milliliters of 0.8% (w/v) piperazine di-acrylamide (PDA) are prepared freshly in 30% (w/v) acrylamide. The mixtures are prepared from stock solutions as indicated in **Table 3**.

**Table 3**
**Acrylamide Solutions**

|                           | 9% Acrylamide mixture | 16.5% Acrylamide mixture |
|---------------------------|-----------------------|--------------------------|
| Acrylamide/PDA solution   | 16.8 mL               | 30.8 mL                  |
| 1.5 $M$ Tris-HCl, pH 8.8  | 14.0 mL               | 14.0 mL                  |
| 5% (w/v) $Na_2SO_3$       | 274 µL                | 274 µL                   |
| 10% (w/v) APS             | 210 µL                | 210 µL                   |
| TEMED                     | 18 µL                 | 18 µL                    |
| $H_2O$                    | 24.6 mL               | 10.6 mL                  |

APS, ammonium persulfate; TEMED, $N,N,N',N'$-tetra methylethylenediamine.

2. Reswelling solution. The solution is prepared freshly prior to use. One hundred twenty five microliters per sample are needed. The composition is: 7 $M$ urea, 2 $M$ thiourea, 2% (w/v) 3-[(3-cholamidopropyl)dimethyl-ammonio]-1-propanesulfonate hydrate (CHAPS), 2% (w/v) 3-[$N,N$-dimethyl (3-myristoylaminopropyl)ammonio]propanesulfonate amidosulfobetaine-14 (ASB-14), 2% (v/v) IPG nonlinear pH 3.0–10.0 buffer, 2 m$M$ tributyl-phosphine (from 1 $M$ stock solution in isopropanol), 65 m$M$ DTT.

3. Reequilibration solution. The solution is prepared freshly prior to use. Twelve milliliters per sample are needed. The composition is 50 m$M$ Tris-HCl, pH 8.8, 4 $M$ urea, 2 $M$ thiourea, 2.5% (w/v) acrylamide, 30% (w/v) glycerol, 2% (w/v) SDS, 5 m$M$ tributyl-phosphine, traces of Bromo phenol Blue *(3)*.

4. Running buffer: 198 m$M$ glycine, 125 m$M$ Tris, 0.1% (w/v) SDS. The buffer is prepared from 10X stock solution. Dilution is prepared prior to use.

5. Agarose solution: 0.5% (w/v) agarose in running buffer containing 2 m$M$ tributyl-phosphine. The solution is prepared freshly prior to use.

6. Colloidal Coomassie staining, solution I: 2% (v/v) $H_3PO_4$, 50% (v/v) methanol *(4)*. The solution is prepared freshly prior to use.

7. Colloidal Coomassie staining, solution II: 2% (v/v) $H_3PO_4$, 34% (v/v) methanol, 17% (w/v) $(NH_4)_2SO_4$ *(4)*. The solution is prepared freshly prior to use.

8. Water-saturated butanol. Combine in a bottle 50 mL of butanol with at least 5 mL of water. Use top phase to overlay gels. Store at room temperature indefinitely.

## 2.6. Devices and Solutions for Autoradiography

1. Devices. Gels are dried on 3MM Chr paper (Whatman International Ltd., Kent, England) using a Bio-Rad gel dryer Model 583. Autoradiography cassettes and BioMax MR-2 film are from Eastman Kodak Co., Rochester, NY.

2. Autoradiography enhancer solution: 125 m$M$ salicyclic acid in 40% methanol.
3. Autoradiograph developer and fixer: 1X GBX developer and 1X GBX fixer are from Eastern Kodac Co.

## 2.7. Software for Image Acquisition and Analysis

The software used for image acquisition and analysis is provided by Amersham Biosciences. Images of autoradiographs are acquired with the Personal Densitometer SI. Radioactive signal relative to each radio-labeled protein is quantified using ImageQuant 5.1. Image Master Elite software v3.10 was used for *in silico* superimposition of gel and autoradiography.

## 3. Methods

## 3.1. Multiprotein Identification in 2D Maps

As already described in the introduction, **Fig. 2** summarizes the steps required for the simultaneous identification of several proteins on the same 2D map. The detailed protocol of each step is reported next.

### 3.1.1. Preparation of Bacterial Proteins (Step 1a) and Chromosomal DNA (Step 1b)

The protein sample used to illustrate the methodology is an outer membrane protein preparation of *Neisseria meningitidis* strain MC58 *(2)*. Briefly, bacteria were grown to confluence on supplemented GC agar plates at 37°C in a humidified atmosphere containing 5% $CO_2$. Bacteria (approx $10^{11}$ cells) were harvested from 10 plates and resuspended in 10 mL of 40 m$M$ Tris. Chromosomal DNA was prepared according to standard procedures *(5)* from 1 mL of cell suspension and stored at a concentration of 1 mg/mL at −20°C. The remaining 9 mL were used for membrane protein preparation. Bacteria were inactivated by 45-min incubation at 65°C, cooled on ice in the presence of benzonase (1000 U), and bacterial cells were disrupted with the French Press apparatus at 18,000 psi. The lysate was centrifuged at 70,000$g$ overnight at 5°C. The pellet was washed twice with 40 m$M$ Tris-base and resuspended in 2 mL of reswelling solution. After centrifugation at 100,000$g$ at 10°C for 3 h, the supernatant containing the solubilized outer membrane proteins was aliquoted and stored at –80°C. Typically, 10 mg of outer membrane proteins were obtained.

### 3.1.2. PCR Amplification and Cloning of the Genes of Interest

TTRs are usually carried out using circular plasmids as templates. Briefly, the genes of interest (*see* **Note 1**) were PCR amplified from *Neisseria meningitidis* strain MC58 chromosomal DNA using Pwo DNA polymerase and appropriate synthetic forward and reverse primers carrying the *Nde*I and *Xho*I (or *Hind*III) restriction sites for the insertion of the gene into the plasmid pET-21b+ (*see* **Subheading 2.2.**). In pET-21b+ a stop codon was introduced upstream from the nucleotide sequence coding for the hexa-histidine tag to avoid the addition of the His-Tag to the C-terminus of the proteins of interest (*see* **Note 2**). Plasmid preparations were carried out using the Qiagen kit (*see* **Subheading 2.2.**), and plasmid DNAs were stored at a concentration of 1 mg/mL at –20°C (*see* **Note 3**).

### 3.1.3. In Vitro TTRs

In vitro-coupled transcription-translations are carried out as described by Pratt *(6)*, with minor modifications. Reactions are performed in 10 µL final volume containing 20–25 µg/mL plasmid DNA or 8–16 µg/mL linear DNA (*see* **Subheading 3.1.2.**), 0.42 m$M$ $^{14}$C-labeled L-leucine and L-lysine, and 4 mg/mL of *E. coli* S30-extract proteins (*see* **Note 4**).

1. For 10 TTRs, add in a microfuge tube 42 nmol of radioactive leucine and lysine.
2. Reduce the volume of the radioactive amino acids to 1–2 µL with a SpeedVac apparatus (Savant, Holbrook, NY).
3. Add 80 µL of reaction mixture into the microfuge tube containing the radioactive amino acids.
4. Mix and split the reaction mixture into 10 TTR tubes (8 µL per tube).
5. Add to each tube the DNA of interest (20–25 µg/mL plasmid or 8–16 µg/mL linear fragment, volume should be about 2 µL).

6. Allow the reaction to proceed for 3 h at 37°C (*see* **Note 5**).
7. Add 90 μL of cold ethanol to each tube and incubate for 30 min at −20°C.
8. Mix and transfer 15 μL of each reaction into new tubes for SDS-PAGE analysis, while the remaining 75 μL are used for 2DE.
9. Recover the proteins by 10 min centrifugation at 13,000*g*.
10. Dry protein pellets with a SpeedVac apparatus (Savant), and solubilize them for SDS-PAGE or 2DE (*see* **Subheadings 3.1.4.**, **steps 1** and **12**, respectively).

## 3.1.4. Quantification and Mixing of TTRs

1. Dissolve the pellets from the 15-μL samples (*see* **Subheading 3.1.3.**, **steps 8–10**) with 5 μL of SDS-PAGE loading buffer.
2. Heat the samples for 3 min at 100°C.
3. Resolve proteins by standard SDS-PAGE using a 12.5% polyacrylamide gel.
4. Soak the gel with the autoradiography enhancer solution (*see* **Subheading 2.6.**).
5. Dry the gel on 3MM Chr paper using the gel dryer (*see* **Subheading 2.6.**).
6. Autoradiograph the gel overnight using BioMax MR-2 film, at −80°C, in the autoradiography cassettes.
7. Develop the film 2–3 min with autoradiograph developer (*see* **Subheading 2.6.**).
8. Wash the autoradiograph with water.
9. Fix the autoradiograph 1 min with autoradiograph fixer (*see* **Subheading 2.6.**).
10. Acquire the image of the film with a Personal Densitometer SI. To guarantee a sufficiently high resolution, we routinely acquire the image at 50 μm per pixel and 12 bits.
11. Quantify the radioactive signal relative to each radio-labeled protein using ImageQuant 5.1 software.
12. Based on quantification, normalize the radioactive signals and mix appropriate aliquots of each TTR (*see* **Note 6**).

## 3.1.5. 2DE of Protein Sample and TTRs

Protein sample and TTRs (*see* **Note 1**) are mixed with reference proteins (*see* **Note 7**). Sample proteins and TTRs are then separated in the first dimension on nonlinear pH 3.0–10.0 (7 cm) IPG strips (*see* **Note 8**), and in the second dimension on 9–16.5% linear gradient polyacrylamide

SDS gels (1.5 mm thick, 7.3 cm high, 8.3 cm wide). Gels are stained with Colloidal Coomassie Blue.

1. Cast two gels using the Mini-Protean II Multiple-Casting Chamber and develop the acrylamide gradient by gravity using the gradient Former.
2. Cover the gels with butanol saturated with water, cover the Mini-Protean II Multiple-Casting Chamber with Parafilm (American National Can™, Menasha, WI) and allow the gels to polymerize overnight.
3. Mix the protein sample (*see* **Subheading 3.1.1.**) with 1 µg of each protein marker and bring the mixture to a final volume of 125 µL with reswelling solution (*see* **Subheading 2.5.**).
4. Mix TTR pool (*see* **Subheading 3.1.4.**, **step 12** and **Note 6**) with 1 µg of each reference protein and bring the mixture to a final volume of 125 µL with reswelling solution.
5. Before gel loading, dissociate possible protein aggregates by subjecting the mixture to three consecutive cycles of 5-min sonication in a sonicator bath and vigorous agitation.
6. Transfer the protein solutions (protein sample and TTR mixture) into two track lanes of the Immobiline Dry-Strip Reswelling Tray.
7. Cover each solution with an IPG strip (gel side down), without trapping air bubbles under the strips.
8. Overlay the strips with 1.5–3 mL of IPG cover fluid (Amersham Biosciences).
9. Allow strip hydration to proceed overnight.
10. Transfer the IPG strips to the IPGphor strip holders.
11. Place filter paper pads at the end of the gel strips.
12. Cover the IPGphor strip holders with IPG cover fluid.
13. Place the IPGphor strip holders on the IPGphor Isoelectric Focusing System, and focalize the proteins absorbed in the strips by applying the following voltage: 150 V for 35 min, 500 V for 35 min, 1000 V for 30 min, 2600 V for 10 min, 3500 V for 15 min, 4200 V for 15 min, and then 5000 V to reach 10 kVh.
14. After isoelectric-focusing, transfer each strip into 15-mL Falcon tubes and equilibrate each strip for 10 min in 6 mL of reequilibration solution.
15. Repeat **step 14**.
16. Remove the butanol saturated with water from the top of the polymerized gels (one for the sample protein and the other for the TTR mixture [*see* **step 1**]).
17. Wash the surface of the gels with water.
18. Cover the gels with 0.5 mL agarose solution.

19. Lay the strips on top of the agarose bed (one per gel).
20. Cover the strips with the agarose solution and allow solidification at room temperature.
21. Place the gels in the Mini-Protean II Cell and run the second dimension in the running buffer by applying 35 mA per gel until the blue fronts reach the end of the gels.
22. Wash the gels for 10 min in $H_2O$.
23. Fix the gels for 60 min in the Colloidal Coomassie staining, solution I.
24. Wash the gels three times for 10 min in $H_2O$.
25. Incubate the gels for 1 h in the Colloidal Coomassie staining, solution II.
26. Add 0.065 g of Coomassie G250 per 100 mL of Colloidal Coomassie staining, solution II.
27. Allow the staining to develop for 48 h.
28. Destain the gels with water until the background staining is removed (usually, three consecutive washes of 10 min each are sufficient).
29. Acquire the image of the 2D gel on which the sample proteins have been resolved using a high-resolution densitometer (*see* **Subheading 3.1.4.**, **step 10**).
30. Soak the gel with the autoradiography enhancer solution.
31. Dry the TTR gel as described in **Subheading 3.1.4.**, **step 5**.

### 3.1.6. Autoradiography of 2DE Gel

1. Fix the dried TTR gel on the bottom of the autoradiography cassette using adhesive paper.
2. In the dark, lay a BioMax MR-2 film on top of the gel and fix it using adhesive paper.
3. Carefully and precisely mark the position of the film with respect to the gel. This step is critical because, after development and fixing, the autoradiograph has to be relocated on the gel exactly in the same position as the one used during autoradiography.
4. Expose the film overnight at –80°C.
5. Develop and fix the film as described in **Subheading 3.1.4.**, **steps 7–9**.
6. Lay the developed, dried film back on the gel in its original position using the markers to properly align the film with the gel.
7. Using black ink, precisely mark the positions of the protein markers on the autoradiograph.
8. Acquire the digital image of the autoradiograph with marked protein markers as described in **Subheading 3.1.4.**, **step 10**.

### 3.1.7. In Silico Superimposition of TTR Film and Sample 2D Gel

For an accurate matching process, a computer program should be used that allows matching the digitalized images using protein markers as landmarks. The rationale is to ask the computer algorithm to correct all possible distortions occurred during protein separations by 2D-electrophoresis by forcing the protein markers to coincide. The result of this process is the artificial creation of a single combined gel in which the autoradiographic spots are positioned in the 2D gel of the protein sample where they would have migrated if the TTRs would have been run on the same gel together with the protein sample. The protein spots of the sample gel, which eventually coincide with, or are in close proximity of, the autoradiographic spots correspond to the proteins encoded by the genes used for the TTRs.

For *in silico* superimposition of gel and autoradiography, we used the software Image Master Elite v3.10 (*see* **Subheading 2.7.** and **Note 9**). An example of matching is given in **Note 10**. For those familiar with this software the steps are as follows:

1. Create a combined gel in which the TTR autoradiograph is superimposed on the sample gel using protein markers as landmarks.
2. Scan the combined gel for protein sample spots coinciding with or located in close proximity to TTR spots, by using increasing vector box size (*see* **Note 10**).
3. The protein spots that are closest to the autoradiographic spots (namely, that require the smallest vector box size to give a match [*see* **Note 10**]), are the proteins with the highest likelihood to be encoded by the genes used for the TTR reactions.

## 3.2. Single Protein Identification in 2D Maps

**Figure 2** summarizes the steps required for the identification of a single specific protein on a 2D map. The detailed protocol for each step is reported next:

1. Bacterial proteins (step 1a) and chromosomal DNA preparations (step 1b). Follow the procedure described in **Subheading 3.1.1.**
2. PCR amplification and cloning of the gene of interest. Follow step 2 described in **Subheading 3.1.2.**

3. In vitro TTR. Follow the procedure described in **Subheading 3.1.3.** with the following modifications:

   a. Add 4.2 nmol of radioactive leucine and lysine to a microfuge tube.

   b. Dry the radioactive amino acids in a SpeedVac apparatus (Savant).

   c. Add 8 μL of reaction mixture to the microfuge tube containing the dried radioactive amino acids and mix.

   d. Add 2 μL of DNA (200–250 ng if circular plasmid is used, or 80–160 ng if linear, amplified fragment is used).

   e. Allow the reaction to proceed for 3 h at 37°C (*see* **Note 5**).

   f. Add 90 μL of cold ethanol and keep at –20°C for 30 min.

   g. Mix and transfer 15 μL into a new tube to be subsequently used for SDS-PAGE analysis. The remaining 75 μL will be used for 2DE.

   h. Precipitate the proteins in both tubes by 10 min centrifugation at 13,000$g$.

   i. Dry protein pellets with a Speed Vac, and solubilize them for SDS-PAGE or 2DE (*see* **Subheadings 3.1.4.**, **steps 1** and **12**, respectively).

## 3.3. Analysis of the TTR by SDS-PAGE

The analysis of the TTR by SDS-PAGE is recommended to define the correct amount of TTR to be subsequently added to the protein sample (*see* **Subheading 3.4.**, **step 3**) to have a visible autoradiographic spot after overnight exposure of the 2D gel.

1. Dissolve the pellet from the 15-μL sample (*see* **Subheading 3.2. step 3, item g**) with 5 μL of SDS-PAGE loading buffer.

2. Heat the sample for 3 min at 100°C.

3. Resolve proteins by standard SDS-PAGE using a 12.5% polyacrylamide gel.

4. Soak the gel with the autoradiography enhancer solution (*see* **Subheading 2.6.**).

5. Dry the gel on 3MM Chr paper using the gel dryer.

6. Autoradiograph the gel overnight using BioMax MR-2 film, at –80°C, in an autoradiography holder.

7. Develop the film 2–3 min with autoradiograph developer.

8. Wash the autoradiograph with water.

9. Fix the autoradiograph 1 min with autoradiograph GBX fixer.

10. Acquire the image of the film with a densitometer such as the Personal Densitometer SI. To guarantee a sufficiently high resolution, we routinely acquire the image at 50 μm per pixel and 12 bits.

11. Quantify the radioactive signal relative to each radio-labeled protein using ImageQuant 5.1. Add to the protein sample (*see* **Subheading 3.4.**, **step 3**) a sufficient amount of TTR to allow the visualization of only the major translation product of the reaction after overnight autoradiographic exposure of the 2D gel. Usually, this is obtained by adding between 5 and 20 μg of TTR (*see* **Note 11** and **Subheading 3.4.**).

## 3.4. 2DE of Protein Sample and TTR Mixture

1. The day before the electrophoretic separation, cast one gel using the Mini-Protean II Multi-Casting Chamber and develop the acrylamide gradient by gravity using the gradient Former.

2. Cover the gel with butanol saturated with water (*see* **Subheading 2.5.**), cover the Mini-Protean II Multiple-Casting Chamber with Parafilm and allow the gels to polymerize overnight.

3. The day after gel casting, add up to 20 μg of TTR (*see* **Subheading 3.2.**, **step 3i** and **Note 12**) to 200 μg of protein sample (*see* **Subheading 3.2.1.**) and bring to a final volume of 125 μL with reswelling solution.

4. Before gel loading, dissociate possible protein aggregates by subjecting the mixture to three consecutive cycles of 5-min sonication in a sonicator bath and vigorous agitation.

5. Transfer the protein solution into a track lane of the Immobiline Dry-Strip Reswelling Tray.

6. Cover the solution with an IPG strip (gel side down), without trapping air bubbles under the strip.

7. Overlay the strip with 1.5–3 mL of IPG cover fluid (Amersham Biosciences).

8. Allow the strip hydration to proceed overnight.

9. Transfer the IPG strip to the IPGphor strip holder.

10. Place filter paper pads at the end of the gel strip.

11. Cover the IPGphor strip holder with IPG cover fluid.

12. Place the IPGphor strip holder on the IPGphor Isoelectric Focusing System, and focalize the proteins absorbed in the strip by applying the following voltage: 150 V for 35 min, 500 V for 35 min, 1000 V for 30 min, 2600 V for 10 min, 3500 V for 15 min, 4200 V for 15 min, and then 5000 V to reach 10 kVh.

13. After isoelectric-focusing, transfer the strip into a 15-mL Falcon tube and equilibrate the strip in 6 mL of reequilibration solution (*see* **Subheading 2.5.**) for 10 min.
14. Repeat **step 13**.
15. Remove the butanol saturated with water from the top of the polymerized gel (*see* **step 1**).
16. Wash the surface of the gel with water.
17. Cover the gel with 0.5 mL agarose solution.
18. Lay the strip on top of the agarose bed.
19. Cover the strip with the agarose solution and allow solidification at room temperature.
20. Place the gels in the Mini-Protean II Cell and run the second dimension in the running buffer by applying 35 mA until the blue front reaches the end of the gels.
21. Wash the gel for 10 min in $H_2O$.
22. Fix the gel for 60 min in the Colloidal Coomassie staining, solution I.
23. Wash the gel three times for 10 min in $H_2O$.
24. Incubate the gel for 1 h in the Colloidal Coomassie staining, solution II.
25. Add 0.065 g of Coomassie G250 per 100 mL of Colloidal Coomassie staining, solution II.
26. Allow the staining to develop for 48 h.
27. Destain the gel with water until the background staining is removed (usually, three consecutive washes of 10 min each are sufficient).
28. Soak the gel with the autoradiography enhancer solution.
29. Dry the gel as described in **Subheading 3.1.4.**, **step 5**.

### 3.5. Autoradiography of the 2DE Gel

1. Fix the dried TTR gel on the bottom of the autoradiography cassette (Eastman Kodak Co.), using adhesive paper. In the dark, lay a BioMax MR-2 film (Eastman Kodak Co.) on top of the gel and fix it using adhesive paper.
2. Carefully and precisely mark the position of the film with respect to the gel. This step is critical because, after development and fixing, the autoradiograph has to be relocated on the gel exactly in the same position as the one used during autoradiography.
3. Expose the film overnight at –80°C.
4. Develop and fix the film as described on **Subheading 3.1.4.**, **steps 7–9**.

5. Lay the developed, dried film back on the gel in its original position using the markers to properly align the film with the gel. The protein spot(s) matching the radioactive spot(s) correspond(s) to the product of the gene used for TTR.

## 4. Notes

1. In the example given in **Fig. 3**, we performed TTRs from six genes selected from the genome of *Neisseria meningitidis* serogroup B (MenB) *(2)*, for which we were interested to know whether the corresponding proteins were present in a MenB outer-membrane preparation. The selected genes encode the proteins NMB1710 (Glutamate dehydrogenase, NADP-specific), NMB1972 (Chaperonin, 60kDa), NMB1936 (ATP synthetase F1, α subunit), NMB1429 (outer membrane protein PorA), NMB2039 (outer membrane protein PIB), and NMB0382 (outer membrane protein class 4).

2. In the case of proteins with leader sequence, the genes were amplified by replacing the leader sequence portion with the methionine start codon. The presence of leader sequences was established by using Psort, available at http://psort.nibb.ac.jp/form.html.

   For example, the two genes NMB1972 and NMB0382 identified in the 2D map of *Neisseria meningitis* (*see* **Note 10** and **Fig. 4**) were amplified using the following primers:

   FOR-NMB197

   5'-AGAATTC<u>CATATG</u>GCAGCAAAAGACG TACAGTTCGGCA-3',

   REV-NMB1972

   5'-ATACCGC<u>TCGAG</u>TCACATCATGCCGCCCAT ACCACCCA-3',

   FOR-NMB0382

   5'-GGAATTC<u>CATATG</u>GGCGAGGCGTCCGTTCAG GGTTACAC-3',

   and REV-NMB0382

   5'-ATACCGC<u>TCGAG</u>TTAGTGTTGGTGATGAT TGTGTGCCGG-3'

   (The *Nde*I and *Xho*I sites used for the insertion of the PCR fragments into the plasmid pET-21b+ (Novagen, Madison, WI) are underlined). Cloning was performed according to standard procedures *(5)*. Plasmid preparations were carried out using the Qiagen kit (Qiagen GmbH, Hilden, Germany), and plasmid DNAs were stored at a concentration of 1 mg/mL at –20°C.

3. TTRs can also be performed using linear DNA fragments as template *(2)*. It has to be pointed out that if DNA linear fragments are used, the addition of extra nucleotides at the extremities of the fragments are required. This

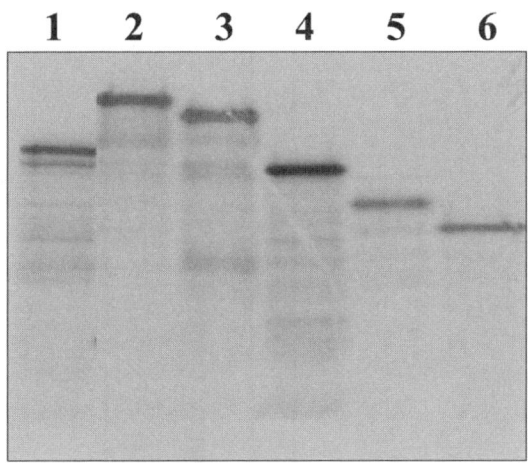

Fig. 3. SDS-PAGE analysis of the products of 6 TTRs. The six *Neisseria meningitidis* genes NMB1710 (lane 1), NMB1972 (lane 2), NMB1936 (lane 3), NMB1429 (lane 4), NMB2039 (lane 5), NMB0382 (lane 6) were amplified and used for in vitro synthesis of the encoded proteins. Six micrograms of each TTR were separated by SDS-PAGE. The gel was then dried and autoradiographed (*see* **Note 6** for details).

is to prevent the T7 promoter and terminator from being degraded by nucleases present in the S30-extract during TTR. For instance, when we included the 78 nucleotides and 103 nucleotides that in pET21b+ precede and follow the T7 promoter and terminator, respectively, we did not observe a substantial impairment of the transcription activities of the fragments during the 3-h incubation.

4. Different S30-extracts are commercially available (e.g., Promega, Madison, WI; Qiagen, Hilden, Germany) and are well suited for the procedure. We prepared the S30-extract in house, using the *E. coli* strain BL21DE3 (Invitrogen, Carlsbad, CA.) containing endogenous T7 RNA polymerase. The S30-extract preparation was performed as described by Pratt *(6)*, adapting the protocol to 2 L of culture. Usually, 8 mL of S30-extract at a protein concentration of 12–15 mg/mL is obtained. When stored at –80°C, this amount of S30-extract is stable for 3–4 mo and is sufficient for more than 2400 TTRs.

5. To enhance protein synthesis, 8 µL of reaction mixture, where the S30-extract is replaced by $H_2O$, can be added to each TTR and the reaction continued for additional 3 h.

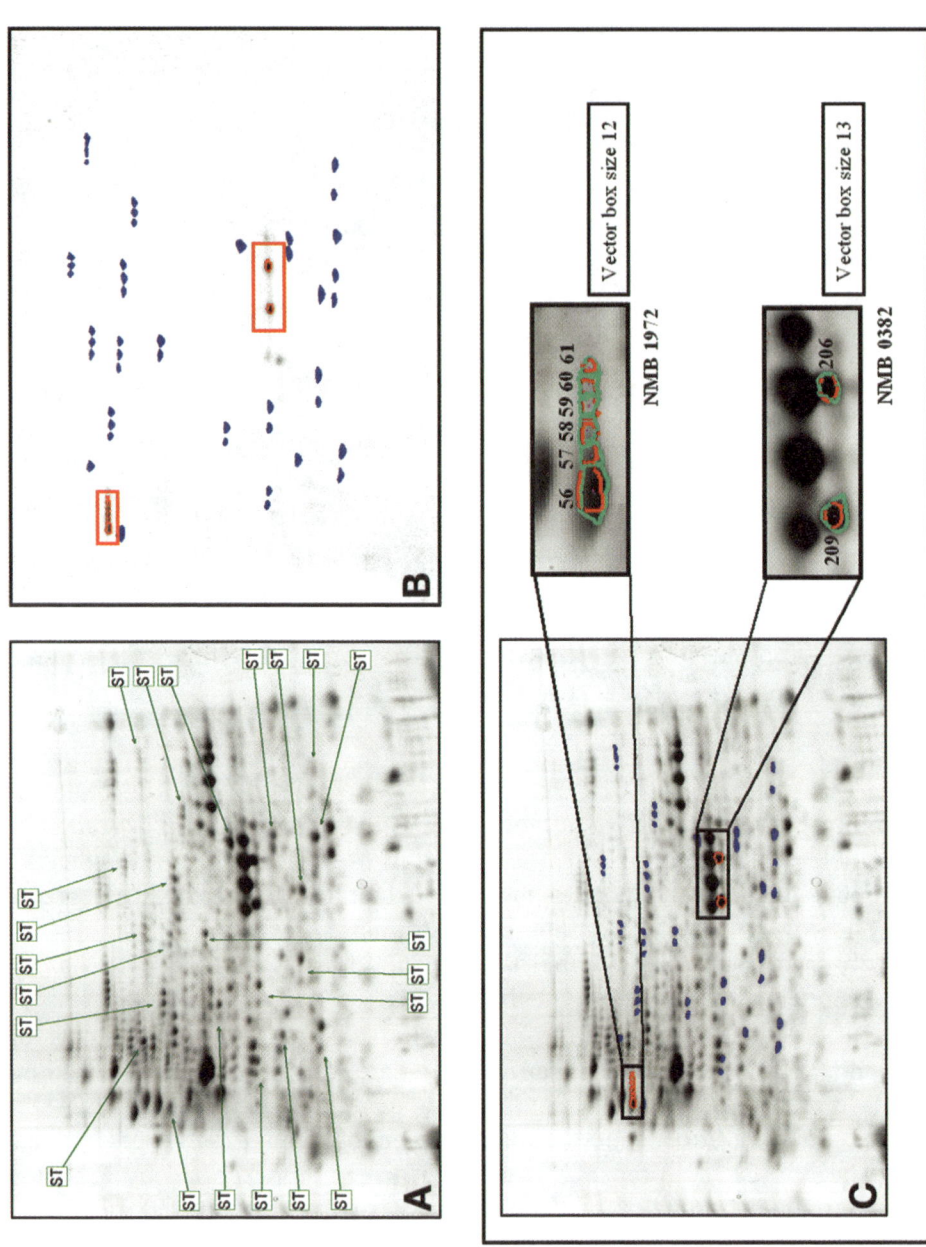

6. It is important that each in vitro translated radioactive protein shows up with similar intensities on the 2D map. To achieve that, normalization by SDS-PAGE analysis is an important step. For instance, **Fig. 3** shows the autoradiograph after 14 h exposure of an SDS-gel on which 15 µL of 6 TTRs are resolved as described in **Subheading 3.1.4.** The scanning analysis performed on the autoradiograph indicated that the relative amounts of the major products of each reaction, normalized with respect to lane 6, were: 1.3, 2.0, 1.5, 2.1, and 0.7 (lanes 1, 2, 3, 4, and 5, respectively). Therefore, if the six TTRs were to be analyzed on the same 2D gel, to visualize only the predominant product of each reaction at the same intensity, the following volumes of each reaction should be mixed: 11.5, 7.5, 10.0, 7.1, 21.4, and 15.0 µL. As an example, we show in **Fig. 4B** the autoradiograph of a 2D gel on which 7.5 and 15 µL of the TTR products of lanes 2 and 6 in **Fig. 3** were separated after mixing. The major reaction products, which resolved in more than one spot as frequently happens in 2DE, showed up at similar intensities (*see* boxed spots), and no intermediate TTR products were visible.

7. Addition of protein markers to both protein samples and TTRs is a fundamental step of the procedure in that the protein markers must be subsequently utilized to properly superimpose the sample gel to the TTR gel. Protein markers should be selected to have an even distribution throughout the gel. This in fact allows the software algorithm to compensate the local gel distortions with higher accuracy. Although commercial kits for 2DE protein markers are available, we made our own set of protein markers consisting in highly purified recombinant proteins with molecular

---

Fig. 4. *(opposite page)* Identification of NMB1972 and NMB0382 proteins in the 2D map of *Neisseria meningitidis* outer membrane proteins. (**A**) Coomassie-Blue staining of the 2D gel of the total *N. meningitidis* outer membrane proteins run together with 22 protein markers (*see* **Table 4**). The protein markers (1µg each) are labeled in green. (**B**) Autoradiograph of the TTRs used for in vitro labeling of NMB1972 and NMB0382 proteins. The two TTRs were separated on a two dimensional (2D) gel together with the same protein markers as in **A** and the gel was autoradiographed. On the autoradiograph the position of the protein markers are labeled in blue. (**C**) Computer-derived image obtained by *in silico* superimposition of the 2D gel image shown in **A** on the TTR autoradiograph shown in **B**. The zoomed boxes indicate the matching of the autoradiographic spots with the gel spots. (For details *see* **Note 10**.)

weights and pIs values ranging from 21.8 to 84.0 kDa, and from 4.89 to 8.62, respectively. The list of the selected proteins is given in **Table 4**.

8. Different IPG strips having broad or narrow pI ranges are commercially available. We recommend the use of broad pI range strips because theoretical and experimental pI values do not always coincide. If a better resolution is needed, narrow-range pI strips can be used once the pI value has been experimentally determined.

9. Analysis with Melanie III has provided comparable satisfactory results.

10. To illustrate the effectiveness of the multi-protein identification procedure, the identification of two meningococcal proteins is reported in **Fig. 4**. In this particular experiment, we were interested in establishing whether the products of the NMB1972 and NMB0382 genes were present in an outer-membrane protein preparation of Group B *N. meningitidis*. **Figure 4A** shows the Coomassie Blue-stained 2D gel on which the membrane proteins were separated together with the mixture of protein markers (1 μg each, labeled in green). **Figure 4B** shows the autoradiograph of a 2D gel (TTR gel), run in parallel with the gel of **Fig. 4A**, in which the two TTRs reactions, carried out using the NMB1972 and NMB0382 genes as templates, were separated in the presence of the same protein markers as the ones used in the gel of **Fig. 4A**. In **Fig. 4B**, the autoradiographic spots corresponding to the radioactive NMB1972 and NMB0382 proteins are boxed in red (*see* **Note 6**), whereas the blue spots correspond to the protein markers whose positions were marked upon superimposition of the autoradiograph on the TTR gel stained with Coomassie Blue. The images of the gel in **Fig. 4A** and of the autoradiograph in **Fig. 4B** were scanned, digitalized and subsequently *in silico*-superimposed using the protein markers as landmarks. **Figure 4C** represents the combined image derived from the *in silico* superimposition of **Fig. 4A,B**. In essence, the computer superimposes the autoradiograph on the protein sample gel using markers as reference spots and in so doing compensates the distortions that occurred during gel running. The result of this process is that the radioactive spots change their position in the virtual combined image according to the corrections. Once proper compensations are done, the computer "scans" around the radioactive spots in search for the nearest protein spots belonging to the protein sample gel and indicates the distance (vector box size) between the center of the radioactive spots and the nearest protein spot. For instance, in **Fig. 3C** the matching between the autoradiographic spots of NMB1972 and NMB0382 (red spots) and the Comassie Blue stained spots (green circles) of the MenB membrane protein gel are zoomed. As shown, the centers of the red and green circles are located only 12 and 13 vector box sizes apart, respectively. Because one box size corresponds to

**Table 4**
**List of Protein Markers**

| Name | Origin | MW KDa | pI |
|------|--------|--------|-----|
| CPn0195 | *Chlamydia pneumoniae* | recombinant GST fusion | 84.0 | 6.78 |
| CPn0197 | *C. pneumoniae* | recombinant GST fusion | 75.0 | 6.29 |
| CPn0278 | *C. pneumoniae* | recombinant GST fusion | 55.0 | 6.67 |
| CPn0324 | *C. pneumoniae* | recombinant GST fusion | 70.1 | 5.22 |
| CPn0336 | *C. pneumoniae* | recombinant GST fusion | 60.2 | 6.21 |
| CPn0420 | *C. pneumoniae* | recombinant GST fusion | 62.8 | 5.65 |
| CPn1064 | *C. pneumoniae* | recombinant GST fusion | 32.0 | 5.00 |
| Glutatione-*S*-Transferase | *Schistosoma japonicum* | | 27.4 | 6.40 |
| CPn0278 | *C. pneumoniae* | recombinant His Tag | 29.6 | 7.01 |
| CPn0764 | *C. pneumoniae* | recombinant His Tag | 47.9 | 7.28 |
| CPn0399 | *C. pneumoniae* | recombinant His Tag | 23.3 | 8.07 |
| CPn1034 | *C. pneumoniae* | recombinant His Tag | 29.0 | 6.05 |
| CPn0113 | *C. pneumoniae* | recombinant His Tag | 41.0 | 5.92 |
| CPn1067 | *C. pneumoniae* | recombinant His Tag | 21.8 | 6.26 |
| CPn0321 | *C. pneumoniae* | recombinant His Tag | 40.9 | 5.57 |
| CPn0273 | *C. pneumoniae* | recombinant His Tag | 22.9 | 6.63 |
| CPn0297 | *C. pneumoniae* | recombinant His Tag | 34.7 | 5.67 |
| CPn0624 | *C. pneumoniae* | recombinant His Tag | 37.7 | 6.60 |
| SAg0681 | *Group B Spreptococcus* | recombinant GST fusion | 65.7 | 8.62 |
| SAg0079 | *Group B Spreptococcus* | recombinant His Tag | 27.5 | 5.41 |
| SAg0628 | *Group B Spreptococcus* | recombinant His Tag | 50.8 | 4.89 |
| NMB1119 | *Neissera meningitidis* | recombinant His Tag | 21.8 | 4.95 |

50 μm in length, practically speaking, the superimposed spots almost co-incide. The conclusion of this experiment was that both NMB1972 and NMB0382 were present in the membrane protein preparation; on the 2D map, the NMB1972 was resolved in six isoforms whereas NMB0382 was constituted by two isoforms.

11. It is very important to load a limited amount of TTR on the gel so that the proteins of the *E. coli* S30-extract do not show up on the gel. This is why the preliminary analysis of TTR using SDS-PAGE and autoradiography is required: it allows to establish the minimum amount of TTR sufficient to visualize the radioactive spots after overnight exposure. In our experience, anything below 20 μg of TTR does not give visible spots under the conditions we used for the Coomassie Blue staining.

12. It has to be considered that in a complex protein mixture, completely unrelated proteins having similar molecular weights and isoelectric points could be present. Because these proteins tend to comigrate on 2D gels, the possibility that the products of the TTRs find a match with an unrelated protein cannot be ruled out. However, on the basis of our experience in proteomic analysis using 2DE coupled to MALDI-TOF, we have estimated that only 1–2% of all visible spots are constituted by more than one protein. Therefore, statistically, only 1 out of 100 protein identifications is expected to be incorrect.

## References

1. Lopez, M. F. (2000) Better approaches to finding the needle in a haystack: optimizing proteome analysis through automation. *Electrophoresis* **21,** 1082–1093.
2. Norais, N., Nogarotto, R., Iacobini, E. T., et al. (2001) Combined automated PCR cloning, in vitro transcription/translation and two-dimensional electrophoresis for bacterial proteome analysis. *Proteomics* **1,** 1378–1389.
3. Herbert, B. R., Molloy, M. P., Gooley, A. A., Walsh, B. J., Bryson, W. G., and Williams, K. L. (1998) Improved protein solubility in two-dimensional electrophoresis using tributyl phosphine as reducing agent. *Electrophoresis* **19,** 845–851.
4. Doherty, N. S., Littman, B. H., Reilly, K., Swindell, A. C., Buss, J. M., and Anderson, N.L. (1998) Analysis of changes in acute-phase plasma proteins in an acute inflammatory response and in rheumatoid arthritis using two-dimensional gel electrophoresis. *Electrophoresis* **19,** 355–363.

5. Sambrook, J., Fritsch, E. F., and Maniatis, T. (1989) *Molecular Cloning: A Laboratory Manual, 2nd ed.* Cold Spring Harbor Laboratory Press, Cold Spring Harbor, New York.

6. Pratt, J. M. (1984) Coupled transcription-translation in prokaryotic cell-free systems, in *Transcription and Translation: A Practical Approach.,* (Hames, B. D. and Higgins, S. J., ed.), IRL Press, New York, pp. 179–209.

# 11

## In Vitro Screen of Bioinformatically Selected *Bacillus anthracis* Vaccine Candidates by Coupled Transcription, Translation, and Immunoprecipitation Analysis

**Orit Gat, Haim Grosfeld, and Avigdor Shafferman**

### Summary

The availability of the *Bacillus anthracis* genome sequence allowed for *in silico* selection of a few hundred open reading frames (ORFs) as putative vaccine candidates. To screen such a vast number of candidate ORFs, without resorting to laborious cloning and protein purification procedures, methods were developed for generation of PCR elements, compatible with in vitro transcription-translation and immunoprecipitation, as well as with their evaluation as DNA vaccines. Protocols will be provided for application of these methods to analyze the anti-*B. anthracis* antibody repertoire of hyperimmune sera or sera from convalescent and from DNA-vaccinated animals.

**Key Words:** In vitro transcription; in vitro translation; reverse vaccinology; vaccine development; *Bacillus anthracis*; bioinformatics; genomics; DNA vaccine; gene gun; immunoprecipitation; serological analysis; immunogenicity screen; antigenicity screen.

From: *Methods in Molecular Biology*, vol. 375,
*In Vitro Transcription and Translation Protocols, Second Edition*
Edited by: G. Grandi © Humana Press Inc., Totowa, NJ

## 1. Introduction

Anthrax is a fatal disease, caused by the aerobic gram-positive spore forming bacterium, *Bacillus anthracis*. The extreme virulence of *B. anthracis* is derived mainly from two key elements: a toxin complex and a capsule *(1)*. Three genes (*pag*A, *cya*, *lef*) encoding for the toxin complex (protective antigen [PA], edema factor [EF], and lethal factor [LF]), are located on a native plasmid, pXO1. PA is the target-cell binding determinant and translocation factor. PA interacts with the two effector moieties: EF, and LF. The complex of PA and EF, named edema toxin, is responsible for extreme tissue edema associated with anthrax, and the complex of PA and LF, lethal toxin, is involved in anthrax-associated death. The genes encoding for functions associated with the synthesis of the poly-γ-D-glutamic acid capsule are located on a second native plasmid, pXO2. The capsule is involved in antiphagocytic activity and survival within macrophages. Although the toxin complex and the capsule are considered the major virulence factors, other proteins encoded by the virulence plasmids and the chromosome, probably work in concert to enable survival and growth of *B. anthracis* in its host, permitting anthrax pathogenesis.

Licensed human vaccines are based on subcellular formulations of secreted bacterial proteins *(2)*. PA is the major constituent of these formulations. Although efficient and safe for human use, the PA-based subcellular vaccines require a prolonged immunization schedule as well as yearly boosting, in order to achieve and maintain full protection. Furthermore, PA vaccines fail to confer protection against some *B. anthracis* strains in specific animal models. All this has led, in recent years, to various attempts to formulate improved alternative anthrax vaccines. The common rational behind these novel approaches is to complement the established protective value of the pXO1-encoded PA, by additional anthrax-derived factors. Accordingly, newly identified *B. anthracis* antigens, encoded by genes located either on the chromosome or on the virulence plasmids, may constitute additive ingredients to PA in order to generate an improved vaccine exhibiting a wider range of protection and requiring a less demanding vaccination regime.

The currently available genomic sequences of numerous human pathogens pave the way for developing novel antibacterial tools. Genome-based search for vaccine candidates in combination with functional genomics studies, prior to protection assessment, was coined "reverse vaccinology" by R. Rappouli *(3)*. Applying "reverse vaccinology" to the genome of a given pathogen entails *in silico* screening for genes, encoding putative surface-exposed or exported proteins. Selected genes are usually cloned and expressed in *Escherichia coli*; the corresponding protein products are then purified and used to immunize mice. Subsequently, protective ability against the disease is assessed. The feasibility of the entire approach relies, therefore, on the availability of a rapid, high-throughput system to evaluate immunity conferred by the bioinformatically selected antigens. Recent completion of the DNA sequence of *B. anthracis* chromosome *(4)*, together with the previously documented sequencing of the two virulence plasmids (*[5]* GenBank accession numbers AF065404 and AF188935), allows for *in silico* analysis of the complete *B. anthracis* genome. A bioinformatic survey, in search for novel vaccine candidate and/or virulence-related genes, was performed on the *B. anthracis* chromosome *(6)* and pXO1 *(7)*. These bioinformatically preselected candidate genes are expressed in vitro, in a cell-free system *(7)*, circumventing the need for a cloning step. By omitting the traditional cloning step in the primary screen, it is possible to screen a large number of genes, without being restricted by efficiency of cloning or possible toxicity of the genes products in a host bacterium. This last aspect is important when dealing with proteins that are predicted to be either membranal or exported, considering the fact that *E. coli*, the commonly used cloning host, does not possess a native secretion apparatus for exporting proteins into the growth medium. The corresponding protein products are now available to be tested for immunoreactivity with antisera produced against live *B. anthracis* strains. This step provides data concerning the immunogenic potential of the respective candidate open reading frames (ORFs) products. The ORF products are further screened by animal vaccination, applying a DNA vaccine-based technique. This step enables the assessment of their ability to elicit an immune response upon administration to animals as single antigens, as well as to generate protective immunity against a lethal challenge.

This chapter concentrates on the in vitro transcription-translation-based procedures involved in the in vitro and in vivo screening steps. An outline of the various stages is provided in **Fig. 1**.

## 2. Materials

### 2.1. PCR Amplification

1. Expand™ High Fidelity system (Roche Molecular Biochemicals).
2. Expand™ Long Template PCR system (Roche Molecular Biochemicals).
3. Reagents for DNA electrophoresis: molecular-biology-grade agarose, TAE electrophoresis buffer, sample-loading buffer, DNA size markers, ethidium bromide stock solution, and distilled deionized water (ddH$_2$O).

### 2.2. Transcription and Translation Reaction and Detection of Protein Products

1. 10 mCi/mL [$^{35}$S]-methionine, specific activity of approx 1200 Ci/mmol.
2. TNT® T7 Quick for PCR DNA kit (Promega).
3. Nuclease-free water.
4. 1X SDS loading dye (Laemmli sample buffer).
5. Reagents for SDS-PAGE electrophoresis: 10–15% sodium dodecyl sulfate (SDS)-polyacrylamide gels, running buffer and [$^{14}$C]-Rainbow size-marker (Amersham).
6. Amplify® Reagent (Amersham).
7. Fixing solution: 7% glacial acetic acid, 20% methanol, 73% ddH$_2$O.
8. Whatman® 3MM filter paper.
9. Kodak X-OMAT® AR film.

### 2.3. Immunoprecipitation Reaction

1. RIPA buffer: 20 m$M$ Tris-HCl, pH 7.5, 150 m$M$ NaCl, 5 m$M$ EDTA, 0.5% sodium deoxycholate, 1% Triton X-100, 0.1% SDS.
2. Specific polyclonal antiserum.

3. Naive serum from the same animal species as the specific polyclonal anti-serum, for background control.
4. TE buffer: 10 m$M$ Tris-HCl, pH 8.0, 0.1 m$M$ EDTA.
5. 1% NaN$_3$.
6. ProteinA-Sepharose 4B Fast Flow (Sigma). The commercially supplied beads contain ethanol, which should be washed away prior to use. Save as 50% slurry in TE buffer containing 0.02% NaN$_3$.

## 2.4. Preparation of Plasmid DNA for Immunization

1. Luria-Bertani (LB) broth: 5 g/L tryptone, 10 g/L yeast extract, 5 g/L NaCl. Sterilize by autoclaving.
2. 100 mg/mL ampicillin stock solution. Sterilize by filtrating.
3. Buffer A: 50 m$M$ Tris-HCl, pH 8.0, 10 m$M$ EDTA, 100 μg/mL RNaseA.
4. Buffer B: 200 m$M$ NaOH, 1% SDS.
5. Buffer C: 3.0 $M$ potassium acetate, pH 5.5.
6. Isopropanol.
7. Cesium chloride (CsCl) ultra pure. Dissolve approx 1 g/mL in ddH$_2$O, density of 1.397.
8. 10 mg/mL Ethidium bromide in ddH$_2$O.
9. *n*-Butanol. Saturate with ddH$_2$O.
10. TE buffer: 10 m$M$ Tris-HCl, pH 8.0, 0.1 m$M$ EDTA.
11. Ethanol absolute and 70% in ddH$_2$O, ice cold. Keep at –20°C.

## 2.5. DNA Immunization With Plasmid DNA Using Gene Gun

1. Helios Gene Gun system (Bio-Rad) contains: Helios Gene Gun and a 9 V battery (sufficient for 1000 shots), helium hose assembly and regulator, Tubing cutter, Tubing Prep Station, gold-coat tubing, cartridges and cartridge holder, 1-μm diameter gold microcarriers, and polyvinylpyrrolidone (PVP).
2. Gold particles (Bio-Rad).
3. Ethanol absolute, unopened bottle. Opened bottle of ethanol absorbs water and the presence of water in the tubing will lead to uneven coating with the gold microcarriers.
4. 20 mg/mL PVP in ethanol, stock solution.
5. 50 m$M$ Spermidine.
6. 1 $M$ CaCl$_2$.

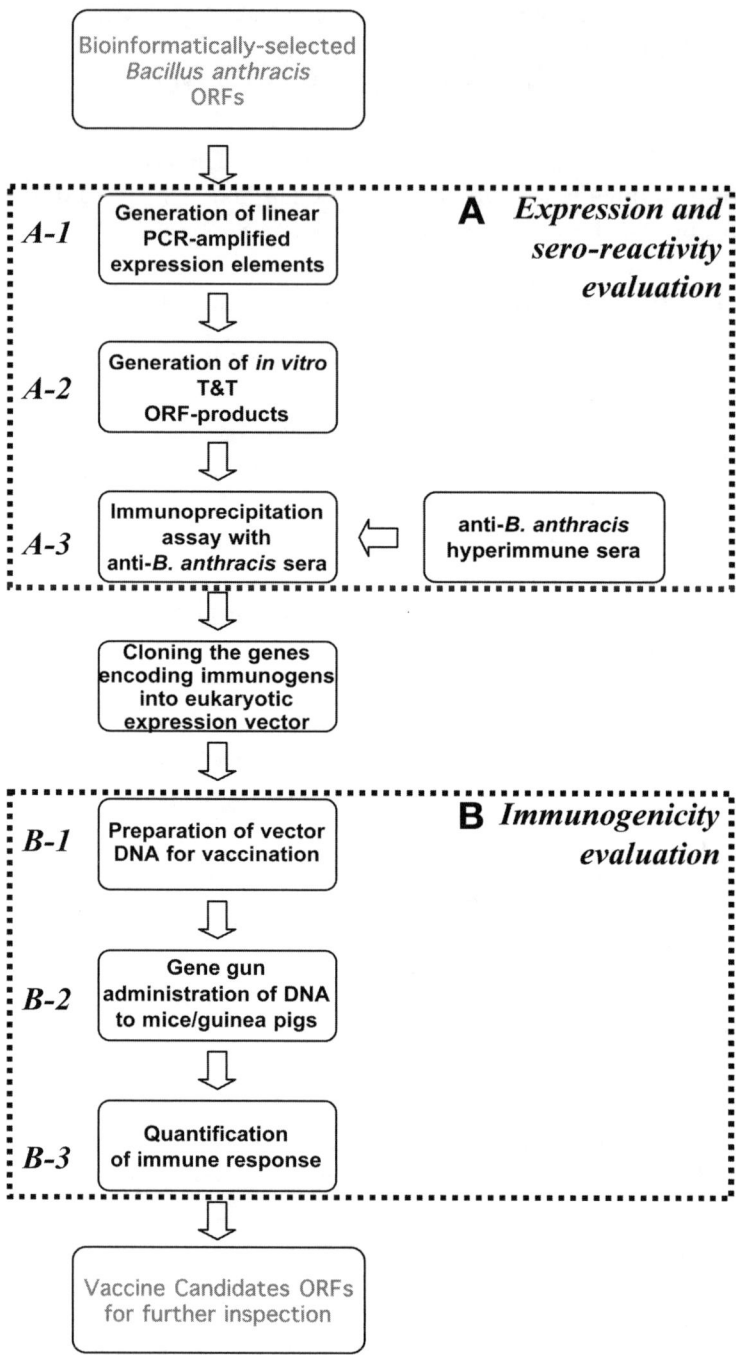

## 2.6. Quantification of Radiolabeled Products Using β-Counter

1. RIPA buffer (*see* **Subheading 2.3.**).
2. Scintillation solution Ecoscint H© (national diagnostics).

## 3. Methods

### 3.1. In Vitro Screen of Candidate ORFs

The in vitro screen includes expression of the selected ORFs through linear PCR expression cassettes (*see* **Subheading 3.1.1.**). The linear PCR expression elements, representing all the bioinformatically selected ORFs, are then subjected to in vitro-coupled transcription and translation (T&T) reaction using rabbit reticulocyte lysate, for generation of each of the protein products individually (*see* **Subheading 3.1.2.**). The T&T reactions are performed in the presence of [$^{35}$S]-methionine, enabling detection of the products via fluorography. Sero-reactivity is monitored by immunoprecipitation (*see* **Subheading 3.1.3.**) of the T&T radioactive products with a series of polyclonal antisera generated in animals recuperated from *B. anthracis* infection (*see* **Note 1**).

---

Fig. 1. (*opposite page*) A general scheme of the described procedures. The bioinformatically selected vaccine candidate *Bacillus anthracis* open reading frams (ORFs) (*see* **refs. 6** and **7** for examples) are subjected to a three-step in vitro screen (*A1*, *A2*, and *A3*, corresponding to **Subheadings 3.1.1.**, **3.1.2.**, and **3.1.3**, respectively). The screen is based on the ability of anti-*B. anthracis* sera to bind in vitro-expressed proteins on linear PCR-produced genes. The genes coding for polypeptides scoring positively in this screen, are subsequently cloned in a eukaryotic expression plasmid vector for administration into experimental animals as DNA vaccine (*B1* and *B2*, corresponding to **Subheadings 3.2.1.** and **3.2.2.**). The immunogenicity of the ORF-product, expressed in vivo following DNA immunization, is evaluated by immunoprecipitation reaction (B3 and **Subheading 3.2.3.**). T&T = coupled in vitro transcription and translation.

The T&T thus serves two purposes: to screen the initial large pool of ORFs for promising vaccine candidates (in the context of *box A*), and to assess the immune response to specific proteins following DNA vaccination (*box B*).

### 3.1.1. Production of Linear PCR Elements

Genomic DNA from *B. anthracis* Vollum strain was extracted using standard method with modifications addressing requirements for gram-positive bacterial lysis *(8)*. The preparation contained chromosomal and plasmid DNA, and was used as the template for PCR reactions.

1. Construct oligonucleotide primers as follows. All the 5' primers should include a common sequence coding for T7 promoter, Kozak consensus sequence and restriction sites, upstream of the specific sequence for amplification of the selected gene (**Fig. 2A**). All the 3' primers contain a stop codon followed by restriction sites (**Fig. 2B**). The restriction sites are compatible with cloning into the eukaryotic expression vector (*see* **Subheading 3.2.1.**): either *Kpn*I or *Sal*I in the 5 primers, and *Sma*I, *Bgl*II, and *Not*I in the downstream primers. Amplicons of the respective ORFs should include the entire coding sequence, excluding only 5' signal peptides (encoding secretion signals). Whenever necessary, removal of $T_M$ segments, highly hydrophobic domains, or common antigenic domains, as well as reduction in size, can be performed (*see* **Note 2a–c**).

2. Primers may be used directly in the PCR reaction, without purification. Apply the Expand High Fidelity (for genes up to 2000-bp long) or the Expand Long Template (for larger genes) PCR systems. Final volume of the PCR reaction is 100 µL.

3. Following amplification, analyze 2–3 µL of the PCR reactions by agarose gel electrophoresis along with DNA size markers, to verify the size of the product and to estimate its quantity. Final DNA concentration should be 50–100 ng/µL.

4. The PCR products may be used directly in the T&T reaction (*see* **Note 3**).

### 3.1.2. In Vitro Transcription and Translation of Linear PCR Elements

The linear PCR elements are translated individually by using a coupled rabbit reticulocyte lysate T&T kit, designed for optimum expression of PCR templates, with T7 RNA polymerase and [$^{35}$S]-methionine. The T&T products are analyzed by SDS-polyacrylamide gel electrophoresis (PAGE) followed by fluorography.

**A**

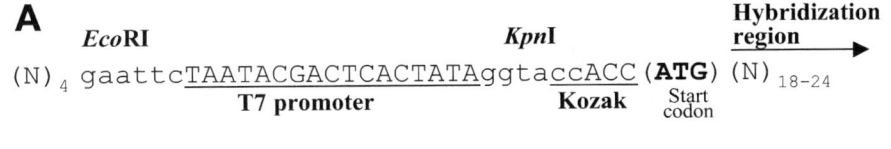

**B**

Fig. 2. Design of 5' (**A**) and 3' (**B**) PCR primers.

1. Thaw the reaction components on ice.
2. Assemble the following reaction components in a 1.5-mL microcentrifuge tube:

   | | |
   |---|---|
   | TNT® T7 Quick Master Mix | 40 μL |
   | [³⁵S]-Methionine | 2 μL |
   | PCR-generated DNA template (100–250 ng) from **Subheading 3.1.1.** | 2–5 μL |
   | Nuclease-free water to a final volume of | 50 μL |

3. Gently mix by pipetting. If necessary, centrifuge briefly.
4. Incubate the reaction at 30°C for 90 min. The reaction is now complete
   and can be stored at –20°C.
5. For performing the SDS-PAGE analysis, remove a 1- to 4-μL aliquot and
   add it to 20 μL SDS loading dye. Heat at 90–100°C for 10 min. Use screw-
   cap tubes to prevent the cap from popping off and releasing the radioactive content.
6. Load about a half of the denatured sample onto a 10–12% SDS-PAGE gel.
   The rest of the sample can be stored at –20°C. Load also a radiolabeled
   size-marker.
7. Carry on with the electrophoresis. It is recommended to stop the run before the bromophanol blue dye front has reached to the bottom of the gel,
   to facilitate disposal of the unincorporated labeled methionine that
   comigrates with the dye.
8. Transfer the polyacrylamide gel to a plastic box containing fixing solution. Shake slowly for at least 4 h at room temperature. Dispose the
   destaining solution to a radioactive waste tank.

9. Cover the gel with Amplify reagent, for fluorographic enhancement of the signal.
10. Place the treated gel on a sheet of Whatman 3MM filter paper, cover the other side of the gel with plastic wrap and dry at 80°C under vacuum in a gel dryer, for at least 1.5 h.
11. Remove the plastic wrap and expose the gel to Kodak X-OMAT® AR film for 4–6 h at –70°C.

### 3.1.3. Immunoprecipitation of T&T Products

Individual $^{35}$S-radiolabeled T&T reaction mixture is reacted with anti-serum and the complex T&T product-antibody is immunoprecipitated (IP reaction) by using protein A-Sepharose beads. Detection of the precipitated T&T product is performed by SDS-PAGE fluorography (*see* **Subheading 3.1.2.**).

1. Remove 2–4 µL aliquot of the T&T reaction (from **Subheading 3.1.2.**) and add it to RIPA buffer to a final volume of 100 µL (*see* **Note 4**).
2. Add 2–10 µL of the tested polyclonal antiserum (*see* **Note 5**). Mix gently. A negative control of naïve serum from the relevant species should be included for each T&T reaction and in every assay.
3. Incubate for 1 h at 30°C, or for overnight at 4°C.
4. Add 50 µL of 1:1 suspension of washed protein A-sepharose beads with a chopped-off pipet tip.
5. Incubate at 4°C for 1 h with slow agitation in an end-over-end shaker.
6. Centrifuge for 1 min to collect the beads at 4°C in a minifuge and discard the supernatant into a radioactive waste tank.
7. Resuspend the beads gently in 1-mL RIPA buffer, centrifuge and discard the supernatant as before, to wash the beads. Take care when aspirating the supernatant to leave a visible distinct liquid layer over the precipitated beads.
8. Repeat the washing step two more times. Following the third wash, resuspend the beads gently in 50 µL SDS loading buffer.
9. Heat the tube at 90–100°C for 10 min.
10. Mix gently, spin down the beads for approx 5 s and transfer the supernatant to a fresh tube.
11. Analyze by SDS-PAGE (*see* **Subheading 3.1.2.**), side-by-side with an equivalent amount of the T&T reaction product used for the IP reaction (in **step 1**).

## 3.2. Direct In Vivo Antigenicity Screen

The PCR amplified fragments coding for positively identified immunogens are directly cloned via a eukaryotic expression plasmid vector, which allows expression of the bacterial genes in a mammalian host following DNA vaccination. Administration of the purified plasmids encoding for the antigens (*see* **Subheading 3.2.1.**) is performed into shaved abdominal skin area of mice or guinea pigs, using the gene-gun route of immunization (*see* **Notes 6** and **7**), as described in **Subheading 3.2.2.** Measurements of antibody titers following DNA-immunization include quantitative immunoprecipitation titration of analytical amounts of the radiolabeled T&T products (*see* **Subheading 3.2.3.**).

### 3.2.1. Preparation of Plasmid DNA for Vaccination

All vector constructs are based on the eukaryotic expression vector, pCI, which carries the eukaryotic cytomegalovirus promoter, a recombinant chimeric intron and the simian virus 40 polyadenylation signal for efficient in vivo expression, in addition to the T7 promoter for in vitro T&T expression. The desired plasmid DNA, bearing recombinant gene of interest (*see* **Subheading 3.1.1.** for PCR amplification of the desired ORFs), is prepared by an alkaline lysis method followed by CsCl gradient centrifugation (*see* **Note 8**).

1. Propagate the desired *E. coli* transformant strain in 500-mL volume of LB medium containing 100 mg/L ampicillin in a 2-L shaking flask at 37°C with vigorous agitation (250 rpm).
2. To harvest bacterial cells, transfer the culture to GSA buckets (200 mL in each). Collect cells by centrifugation in a Sorvall® at 5000 rpm (4000*g*) for 15 min, at room temperature.
3. Resuspend by extensive up-and-down pipetting and combine the cell pellets of the 500 mL culture in 25 mL buffer A. Add 25 mL buffer B and mix gently by inverting the bucket several times. Confirm cell lysis by verifying that the suspension becomes translucent.
4. Add 25 mL buffer C and mix gently, as in **step 3**. A flocculent white precipitate consisting of chromosomal DNA, high-molecular-weight RNA, and potassium/SDS/protein/cell wall complexes will form.

5. Centrifuge at 9000 rpm (10,000g) for 25 min in GSA rotor, and transfer the supernatant to a clean GSA bucket.
6. Determine the volume and add 0.6 vol of isoporopanol. Mix well by inverting the tightly closed bucket several times.
7. Centrifuge again at 9000 rpm (10,000g) for 25 min and discard the supernatant. The DNA pellet is almost invisible at this step.
8. Dry the translucent DNA pellet in the open, inverted bucket at room temperature, for about 15 min.
9. Resuspend the dried DNA pellet in 13.2 mL CsCl solution and transfer to a Quick-Seal polyallomer tube, suitable for centrifugation in an ultracentrifuge rotor. Add 120 µL of 10 mg/mL ethidium bromide and seal the tube. Mix the content well by inverting the tubes three times. This step and subsequent steps (up to **step 15**) should be carried out in a dimly lit room.
10. Centrifuge the density gradients at 20°C for 24 h at 56,000 rpm (160,000g) for Beckman 90Ti rotor or as appropriate for the rotor. At the end of the run, carefully remove each tube and place it in a test tube rack.
11. Mount one tube in a clamp attached to a ring stand. Two bands of DNA, located in the center of the gradient, should be visible in ordinary light. The upper band consists of linear chromosomal DNA and nicked circular plasmid DNA. The lower more abundant band consists of closed circular plasmid DNA.
12. Collect the band of the supercoiled plasmid DNA slowly (about 2.5-mL volume), using an 18-gage hypodermic needle attached to a 5-mL disposable syringe, taking care not to disturb the upper band. Use a second needle puncture at the top of the tube to avoid vacuum formation. Transfer to a 50-mL disposable polypropylene centrifuge tube with a screw cap.
13. Add three volumes of TE buffer and an equal volume of saturated *n*-butanol, and mix by vortexing. Let the tube stand at room temperature until the organic and aqueous phases have separated (about 5 min).
14. Dispose the upper (organic) phase containing ethidium bromide by transferring it to a waste container.
15. Repeat the extraction step four more times, until all of the ethidium bromide is removed from both phases. From this point on, all steps can be performed in light.
16. Add 2 vol of ethanol absolute, mix and place at 4°C for 1 h.
17. Centrifuge at 3300 rpm (2000g) in a Sigma NR11133 rotor, for 20 min at 4°C, to collect the DNA pellet.
18. Wash the pellet once with 70% ice-cold ethanol. Dispose the ethanol, dry the pellet shortly by vacuum and resuspend in 3-mL volume of TE.

19. Determine DNA concentration and purity (by measuring the optical density at 260/280 nm) and bring to 1 mg/mL in TE buffer (*see* **Note 9**). Keep DNA frozen at −20°C until use.

### 3.2.2. DNA Immunization by Biolistics

Plasmid DNA is introduced to animals by a physical means, using a gene gun to perform particle bombardment. Using this method, high density, subcellular sized particles are accelerated to high velocity to transfect DNA into cells. This is performed with the Helios Gene Gun System (Bio-Rad). Prior to transfection, the plasmid DNAs are precipitated onto 1-μm gold particles (microcarrier), in the presence of spermidine and CaCl$_2$ (*see* **Subheading 3.2.2.1.**). Using a Tubing Prep Station (supplemented with the system), the DNA/microcarrier solution is coated onto the inner wall of Gold-Coat Tubing, which is then cut into cartridges, by using the Tubing Cutter (**Subheading 3.2.2.2.**). The cartridges are now ready to be used with the Helios Gene Gun for injecting the DNA into the epidermis cells of the experimental animals (**Subheading 3.2.2.3.**). The bombardment is performed on bare skin, in the abdominal region (**Subheading 3.2.2.4.**).

### 3.2.2.1. COATING THE MICROCARRIERS WITH DNA

1. Prepare fresh PVP working solution, by diluting the stock solution to a concentration of 20 μg/mL with ethanol. Prepare 3.5 mL for each 30 mm length of Gold-Coat tubing. Keep the solutions tightly capped.
2. Weight 20 mg gold microcarriers into a 1.5-mL microfuge tube.
3. Add 50 μL of 0.05 *M* spermidine to the gold microcarriers. Vortex for 1 min, to break up gold clumps.
4. Add 50 μL (50 μg) plasmid DNA (from **Subheading 3.2.1.**). For cotransfection of two or more plasmids, add the plasmids mixture at this step, keeping the final volume of 50 μL (*see* **Note 10**).
5. Add 50 μL (equal to the volume of spermidine added in **step 3**) of 1 *M* CaCl$_2$ dropwise to the mixture, with vigorous vortexing.
6. Allow the mixture to precipitate at room temperature for 10 min.
7. Spin for 15 s in maximum speed, to pellet the DNA-coated gold, and discard as much as possible from the supernatant.
8. Wash the gold microcarriers three times with 1 mL of fresh ethanol.

9. Resuspend the washed pellet in 500 µL of PVP working solution (from **step 1**). Transfer this suspension to a 15-mL disposable polypropylene centrifuge tube with a screw cap. Rinse the microfuge tube once more with 500 µL of the PVP working solution and add to the polypropylene tube. Add PVP working solution to bring the suspension to final volume of 3 mL.

10. The DNA/microcarrier suspension is now ready for coating the inside of the cartridge tubing. Alternatively, it can be stored (tightly capped) for up to 2 mo at –20°C. Following storage, allow the gold particles suspension to reach room temperature prior to decapping the tube, to avoid water absorption by the cold ethanol.

### 3.2.2.2. LOADING THE DNA/MICROCARRIER SUSPENSION INTO GOLD-COAT TUBING USING THE TUBING PREP STATION

1. Set up a peristaltic pump at a rate of 3.6–7.2 mL/min, to be used later for pumping out ethanol from the tubing. The peristaltic pump should be equipped with a tubing that will tightly fit onto the Gold-Coat Tubing.

2. Set up the Tubing Prep Station and connect it to a nitrogen source. Purge the Gold-Coat Tubing with nitrogen, to ensure it is completely dry. Shut off the nitrogen supply.

3. Insert a beveled uncut piece of tubing into the opening of the Tubing Prep Station. Push the tubing 1 cm beyond the *O*-ring installed in the station.

4. Turn on the nitrogen and adjust the flow to 0.3–0.4 L/min (LPM). Allow nitrogen to flow through for at least 15 min prior to using the Gold-Coat Tubing in the following steps.

5. Remove the Gold-Coat Tubing from the Tubing Prep Station and turn off the nitrogen flow. From the dried Gold-Coat Tubing cut an approx 75-cm length for each 3 mL sample of DNA/microcarrier suspension. Attach a 10-mL syringe to one end of the Gold-Coat Tubing, using adaptor.

6. Suspend the DNA coated microcarrier preparation, by vortexing and inverting the tube (to achieve an even suspension). Quickly draw the gold suspension into the Gold-Coat Tubing. Be careful not to draw air bubbles. Continue drawing the suspension into the tubing to leave about 5 cm of empty tubing at each end.

7. Immediately bring the Gold-Coat Tubing to a horizontal position and slide the loaded tube (with the still attached syringe to its other end) into the groove in the Tubing Prep Station, until it passes through the *O*-ring.

8. Allow the microcarriers to settle within the Gold-Coat Tubing for 3–5 min. Detach the syringe and attach the tubing end to the peristaltic pump

and remove the ethanol solution at a rate of 3.6–7.2 mL/min (30–45 s). Detach the peristaltic pump.

9. Immediately turn the Gold-Coat Tubing 180° while it is in the groove and allow the gold to begin coating the inside surface of the tubing for 3–4 s.

10. Start rotating the Tubing Prep Station. Allow the gold to spread in the tube for 20–30 s then open the flow of nitrogen to dry the tubing while rotating. Continue drying for 3–5 min.

11. Turn off the Tubing Prep Station and the nitrogen, and remove the coated tubing.

12. Prepare 0.5-in. length cartridges using the Tubing Cutter as follows (this procedure requires only a few minutes and should be performed as soon after coating as possible):

   a. Examine the coat on the Gold-Coat Tubing to verify that the microcarriers are evenly distributed over the length of the tubing. Ideally, the gold should be spread uniformly over the entire inside surface of the tubing; however, while drying, it may polarize to one side of the tubing. As long as there are no clumps or bare sections, the tubing can be used for cartridges. With a marking pen, mark any section of tubing to be discarded.

   b. Using scissors, cut off and discard unevenly coated tubing from the end.

   c. Using the Tubing Cutter cut the remaining tubing into 0.5-in. pieces. The pieces will drop into a storage vial.

   d. Cap the vial tightly, wrap with Parafilm and store at 4°C.

   e. It is important to store the coated tubes in a desiccated environment. Tubes stored at 4°C are stable for at least 8 mo.

### 3.2.2.3. OPERATING THE HELIOS GENE GUN

1. A day prior to experiment, check helium supply (50 psi in excess of desired delivery pressure).

2. At day of experiment, connect the Helios Gene Gun to a dry Helium Source, using the helium pressure regulator provided. Turn on the helium flow at 300 psi. While operating the Helios Gene Gun, use ear protection.

3. Activate the Gene Gun with an empty cartridge in place, and make two to three "preshots."

4. Load cartridges into cartridge holder (up to 12 cartridges). Loading only one type of DNA is preferred, to avoid possible confusion. Insert the cartridge holder into the Helios Gene Gun.

5. To deliver the DNA to target tissue (bombard), touch the target tissue with the Gene Gun while the device is perpendicular to the target surface, activate the safety interlock switch and press the trigger button. Release the advance lever. The Gene Gun is ready to deliver the next cartridge.
6. After the bombardment: (1) remove cartridge holder from Gene Gun, (2) remove empty cartridges from cartridge holder, (3) turn off the helium pressure, (4) depressurize the system by turning the regulator valve counterclockwise, and (5) disconnect the Gene Gun from helium source.

#### 3.2.2.4 IN VIVO DELIVERY TO EPIDERMIS

1. Clip or shave the fur off the animals' abdomen (target site).
2. Clean the target site with 70% ethanol and dry.
3. Bombard the DNA-coated gold microcarriers. Vaccination protocol includes three to four immunizations at 2-wk intervals. Each immunization is comprised of two bombardment sites in mice (total of 0.5–1 µg plasmid DNA) and four in guinea pigs (1–2 µg DNA).
4. Discomfort following Helios Gene Gun bombardment of the skin is minimal. The animals may exhibit a transient erythema at the treatment site.
5. Bleed animals, for serum collection, from tail (mice) or by heart puncture (guinea pigs), 2 wk following the last immunization. Following gene gun vaccination, the antibody titer level will last for months.

### 3.2.3. Evaluation of Immune Response by Quantitative Immunoprecipitation of the T&T Products

Accurate quantification of immune response is necessary for (1) characterization of the sera, which is needed as the screening tool and (2) determination of the immune response elicited in the experimental animals upon vaccination. Quantification of response to DNA vaccines relies on immunoprecipitation of constant amounts of $^{35}$[S]-labeled T&T products with serial dilutions of the antiserum from the vaccinated animals.

1. Dilute 10 µL of the $^{35}$S-radiolabeled T&T reaction mixture (from **Subheading 3.1.2.**) into 550-µL volume of RIPA buffer. Mix by pipetting. The volume of the T&T mixture may change according to the efficiency of the T&T reaction. A volume that is equivalent to approx 10,000 protein-incorporated counts per minute is appropriate.

2. Set a series of marked 1.5-mL sterile centrifuge tubes for antiserum dilutions. To the first tube transfer 135 µL of the diluted [$^{35}$S]-radiolabeled T&T reaction mixture (from **step 1**), and to the other tubes add 100 µL.

3. Make an initial 1:10 dilution of the desired antiserum (from **Subheading 3.2.2.4., step 5**) in RIPA buffer, and add 15 µL into the first tube from **step 2** (we recommend starting with 1:100 dilution). Mix by pipetting and transfer 50 µL to the second tube (this will generate a 1:3 dilution).

4. Continue performing sequential 1:3 dilutions by transferring 50 µL from each tube to the next. Mix well after each transfer. This process will generate serial dilutions from the initial 1:100 dilution to 1:300, 1:900, 1:2700, and so on.

5. Prepare a background control by adding 10 µL of 1:10 naive serum to 100 µL of the diluted [$^{35}$S]-radiolabeled T&T reaction mixture (from **step 1**).

6. Incubate the tubes for 1 h at 30°C or overnight at 4°C.

7. Add 50 µL of washed protein A-sepharose beads (in TE buffer) to each tube and continue with the IP protocol (as in **Subheading 3.1.3., steps 4–10**).

8. Transfer 5 µL of each supernatant into 5-mL volume of scintillation solution and count in a β-counter, according to the gating specifications of $^{35}$S. The final serum dilution, which generates twice the count of the background control (from **step 5**), represents the IP titer of the examined sera against the relevant T&T product (*see* **Note 11, Fig. 3A**, and **Note 12**).

9. Alternatively/in addition, analyze 10-µL of the supernatants, side-by-side with the same amount of the T&T reaction (from **step 1**), on SDS-PAGE fluorography, from **Subheading 3.1.2., steps 5–10**. The final dilution that generates a specific band, which correlates with the size of the T&T products, is the IP titer. This dilution should be identical to the result from **step 8** (*see* **Note 11, Fig. 3B**).

## 4. Notes

1. We have recently demonstrated the effectiveness of this approach in the analysis of 11 ORF candidates selected from pXO1 *(7)*. Notably, none of the selected candidate ORF products was reported previously to be expressed by the bacterium, or to represent a true gene. Of the 11 candidate ORFs, 9 were successfully expressed in vitro. The only two genes that were not expressed in the T&T system encode membrane proteins and contain transmembranal ($T_M$) segments dispersed throughout the primary sequence (pXO$_1$-85, a CAAX amino terminal endopeptidase with three

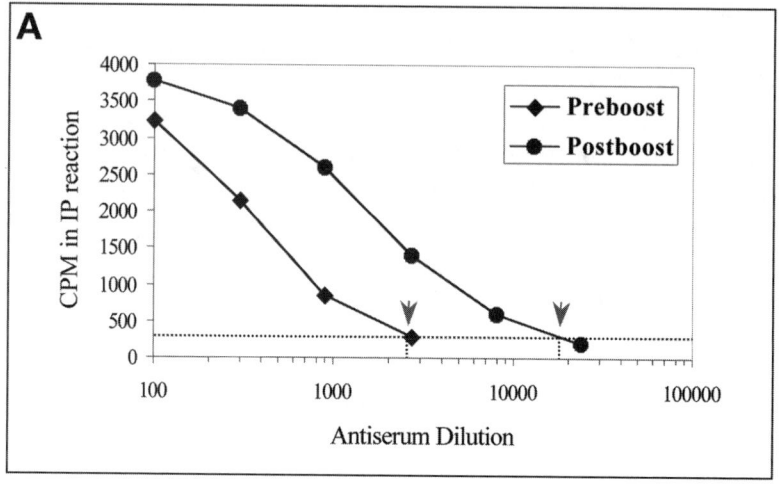

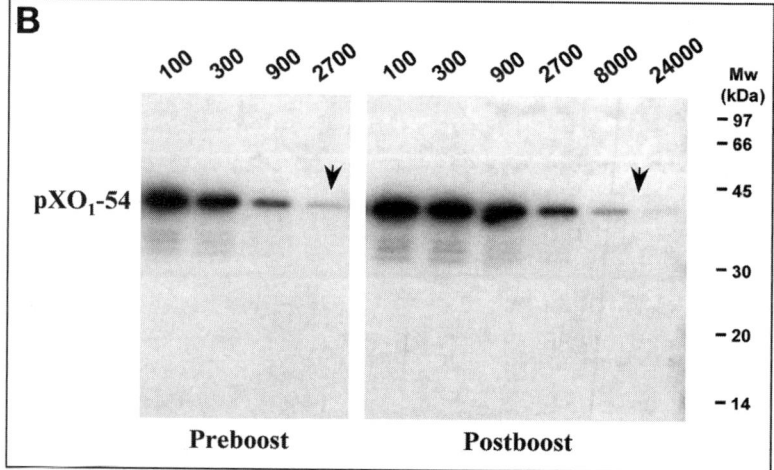

Fig. 3. Evaluation of the immune response in a mouse vaccinated with pCI carrying the pXO$_1$-54 open reading frame (ORF). Sera collected 2 wk following the third gene gun bombardment (**preboost**) and following the forth gene gun bombardment (**postboost**) were compared. (**A**) Five-microliter samples of the serial IP reactions of the titration were counted. (**B**) Ten microliters samples of the serial IP reactions of the titration were resolved on a 12% sodium dodecyl sulfate/polyacrylamide gel electrophoresis gel followed by fluorography. Note that the calculated IP titers (indicated by arrows) by both methods are similar.

T$_M$ segments; pXO$_1$-91, a membrane-associated phospholipid phosphatase with seven T$_M$ segments [*see* **ref. 7** for ORFs' characterization]).

2. To extend the screen, ORFs that could not be expressed efficiently in vitro may be manipulated to yield shorter in vitro expressible derivatives. The following three examples demonstrate handling of such complicated proteins:

    a. Removal of a hydrophobic signal anchor domains. Selected ORFs might code for hydrophobic regions, probably $T_M$ segment. For example, $pXO_1$-79 codes for a 1222-amino acid long, alanine- and serine-rich complex polypeptide, carrying six $T_M$ segments in addition to an N-terminal signal peptide, as well as several repeat domains and an LPXTG sortase-dependent anchor (**Fig. 4A**). Therefore, it appears to be a hydrophobic cell wall protein. Although this very large protein product could be expressed using the T&T system, it could not be resolved by SDS-PAGE, probably due to its high hydrophobicity, which favors the formation of aggregates (**Fig. 4B,Left lane**). In order to overcome these problems, the DNA sequence coding for the N-terminal transmembrane stretch was deleted, resulting in a construct directing the expression of a truncated product ($pXO_1$-79tr, residues 423–1042). The efficient generation of a polypeptide of the expected size was demonstrated (**Fig. 4B,Right lane**).

    b. Generation of fragments of reduced molecular weight from large proteins. The versatility of the T&T-based system is also manifested by the ability to efficiently express shorter derivatives of large proteins, which are more amenable to subsequent analysis. This was done for $pXO_1$-13, a 1320-amino acids putative membrane-anchored multidomain protein (*see* **ref. 7**). Two truncation products, which provide extensive coverage of this ORF-encoded polypeptide, were designed on the basis of domain analysis and indeed, found to be successfully expressed in vitro (**Fig. 5**).

    c. Splicing out competitive antigenic domains. $pXO_1$-54 and $pXO_1$-90 harbor a potentially immunodominant domain common to several proteins, an SLH (S-Layer Homology) domain. Because the SLH-anchorage modality maintain a widespread representation in the proteome of *B. anthracis* (**9**), it was important to determine whether the detected immunoreactivity of these two ORF products is a result of the common SLH domain or specific to each of the unique regions. The sequence encoding the N-terminal SLH domains was thus removed, and the immunogenicity of the specific domains of both $pXO_1$-54 and $pXO_1$-90 was directly demonstrated (*see* **ref. 7**).

3. Optional: the PCR product may be purified prior to the T&T reaction, by using the Wizard PCR Preps purification system (Promega).

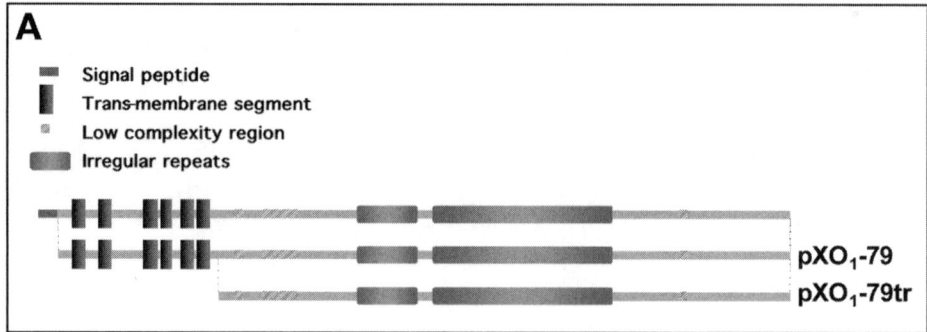

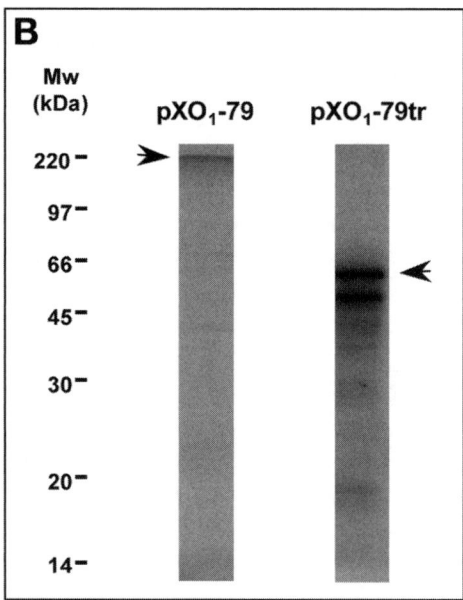

Fig. 4. Ability of the direct in vitro PCR/T&T system to express hydrophobic protein products. (**A**) Schematic representation of $pXO_1$-79 showing full size (130 kDa) and truncated $pXO_1$-79tr (65 kDa) polypeptides. (**B**) The indicated [$^{35}$S]-methionine-labeled products were obtained using PCR-derived DNA template in a T&T reaction and the labeled proteins were analyzed on sodium dodecyl sulfate/polyacrylamide gel electrophoresis following fluorography (*see* **Subheading 3.1.2.**).

4. Theoretically, a mixture of different T&T-products of clearly distinct molecular weights could be immunoprecipitated in a single assay.
5. The anti-*B. anthracis* sera constitute a major screening tool. According to our previous results *(7)*, expanding the repertoire of sera will facilitate a

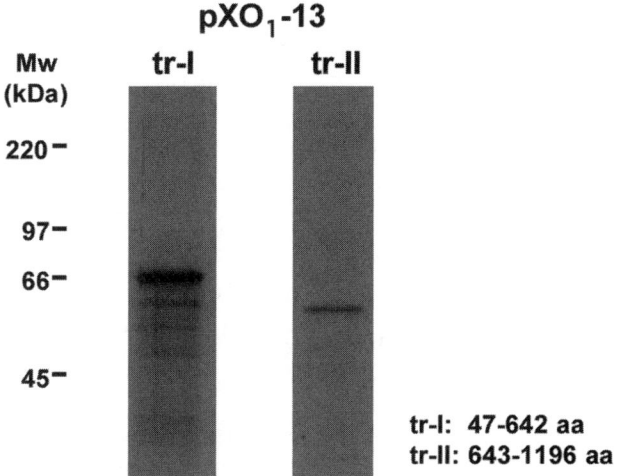

Fig. 5. Generation of truncated product from a very large open reading frame (ORF). The two shortened versions of $pXO_1$-13 represent different parts of the molecule. The full-length product could not be obtained using the T&T system.

more exhaustive identification of immunogens among the bioinformatically preselected candidates. Expression pattern of specific bacterial antigens could differ with the animals used, the mode of infection, the genetic background of the bacterial strain, the level of attenuation of the bacteria used for infection, and the treatment by which the survival of the animals is attained. Relevant sera could be either hyperimmune antisera obtained by multiple administrations of attenuated *B. anthracis* strains, or convalescent antisera obtained by challenging the animal with a fully virulent strain, followed by treatment with antibiotics to attain recovery *(10)*. Recovered animals are then rechallenged with the virulent strain to verify protective immunity and to further potentiate the sera.

6. Direct introduction of the linear PCR expression elements into animals is also possible, yet the use of the plasmid vectors is advantageous for two reasons: (1) we have found that vaccination with linear fragments exhibited reduced reproducibility with respect to the antibody titers obtained, and (2) the amount of DNA generated by PCR is limited, and this may represent a limitation for repeated immunization schedules.

7. To evaluate the DNA delivery route for efficient expression and immunization by the DNA-encoded antigens, we applied two models of DNA administrations: needle injection (either intradermal or intramuscular) and the gene gun intradermal bombardment (*see* **ref. *11*** for comparison

between the two methods). The gene gun-mediated vaccination was by far more effective than needle injection, as established by higher levels of antibody titers achieved with lesser amounts of plasmid DNA. About 200-μg of DNA were required for each needle injection vaccination, as compared to 1–2 μg per gene gun shot. Furthermore, the immune response was characterized by two orders of magnitude higher antibody titers.

8. To our experience, the CsCl gradient purified DNA gave the most reproducible results in the immunization experiments. This became quite crucial in cases where the gene product was of marginal immunogenicity.

9. We routinely obtain approx 5 mg of plasmid DNA from 500 mL of bacterial culture and the DNA has an $A_{260}/A_{280}$ of 1.8 or higher, 80% of which is in the supercoiled form.

10. When introducing a mixture of two different plasmids by gene gun, the overall amount of DNA-coated gold microcarriers should be doubled, by doubling the number of bombardments at different skin sites per one immunization. Reducing the overall amount of each plasmid in the mixture may be at the expense of the efficiency of expression of each plasmid.

11. In the absence of purified protein antigens to be used in ELISA assay, quantification of the immune response is represented as immunoprecipitation (IP) titers. The IP titer can be calculated as the endpoint of IP titration with serial serum dilutions, by one of two means, as detailed in **Subheading 3.2.3.** (1) monitoring the label in a β-counter (**Fig. 3A**) and/or (2) visualization of the IP products in SDS-PAGE fluorography (**Fig. 3B**).

12. When determining the titers by radioactivity monitoring using the β-counter, great care should be taken during the washing step (**Subheading 3.2.3., step 7**), to ensure elimination of the unincorporated [$^{35}$S]-methionine.

# References

1. Mock, M. and Fouet, A. (2001) Anthrax. *Annu. Rev. Microbiol.* **55,** 647–671.
2. Friedlander, A. M., Welkos, S. L., and Ivins, B. E. (2002) Anthrax vaccines. *Curr. Top. Microbiol. Immunol.* **271,** 33–60.
3. Adu-Bobie, J., Capecchi, B., Serruto, D., Rappuoli, R., and Pizza, M. (2002) Two years into reverse vaccinology. *Vaccine* **21,** 605–610.
4. Read, T. D., Peterson, S. N., Tourasse, N., et al. (2003) The genome sequence of *Bacillus anthracis* Ames and comparison to closely related bacteria. *Nature* **423,** 81–86.

5. Okinaka, R. T., Cloud, K., Hampton, O., et al. (1999) Sequence and organization of pXO1, the large *Bacillus anthracis* plasmid harboring the anthrax toxin genes. *J. Bacteriol.* **181,** 6509–6515.

6. Ariel, N., Zvi, A., Makarova, K. S., et al. (2003) Genome-based bioinformatic selection of chromosomal *Bacillus anthracis* putative vaccine candidates coupled with proteomic identification of surface-associated antigens. *Infect. Immun.* **71,** 4563–4579.

7. Ariel, N., Zvi, A., Grosfeld, H., et al. (2002) Search for potential vaccine candidate open reading frames in the *Bacillus anthracis* virulence plasmid pXO1: in silico and in vitro screening. *Infect. Immun.* **70,** 6817–6827.

8. Johnson, J. L. (1994) Similarity analysis of DNAs, in *Methods for General and Molecular Bacteriology,* (Gerhardt, P., Murray, R. G. E., Wood, W. A., and Krieg, N. R., eds.), ASM, Washington, DC, pp. 655–682.

9. Chitlaru, T., Ariel, N., Zvi, A., et al. (2004) Identification of chromosomally encoded membranal polypeptides of *Bacillus anthracis* by a proteomic analysis: Prevalence of proteins containing S-layer homology domains. *Proteomics* **4,** 677–691.

10. Altboum, Z., Gozes, Y., Barnea, A., Pass, A., White, M., and Kobiler, D. (2002) Postexposure prophylaxis against anthrax: evaluation of various treatment regimens in intranasally infected guinea pigs. *Infect. Immun.* **70,** 6231–6241.

11. Grosfeld, H., Cohen, S., Bino, T., et al. (2003) Effective protective immunity to *Yersinia pestis* infection conferred by DNA vaccine coding for derivatives of the F1 capsular antigen. *Infect. Immun.* **71,** 374–383.

# 12

## Functional Expression of Type 1 Rat GABA Transporter in Microinjected *Xenopus laevis* Oocytes

**Stefano Giovannardi, Andrea Soragna, Simona Magagnin, and Laura Faravelli**

### Summary

In this chapter we describe technical aspects and experimental potential of the two electrodes voltage clamp (TEVC) electrophysiological approach applied to the *Xenopus* oocyte-expression system.

This technique is addressed to the study of a particular class of expressed proteins, those responsible to drive ion fluxes through the plasma membrane. In fact the voltage-clamp technique provides the most direct and sensitive measurement of the functional properties of ion channels and electrogenic transporters, allowing specific ion currents to be recorded under well-defined voltage conditions and temporal control.

Besides the study of the physiological properties of specific ion channels as well as their pharmacological modulation, further applications of the TEVC on oocytes include the possibility to introduce single point mutations in the channel construct and to infer to possible structural aspects and functional involvements of single amino acidic residues. To achieve these results these technique should be strictly tied to basic molecular biology techniques. Recent advance of this technique in drug discovery procedures have been briefly enlightened.

**Key Words:** *Xenopus laevis* oocytes; functional expression; cRNA microinjection; two electrodes voltage clamp; high-throughput screening.

From: *Methods in Molecular Biology,* vol. 375,
*In Vitro Transcription and Translation Protocols, Second Edition*
Edited by: G. Grandi © Humana Press Inc., Totowa, NJ

## 1. Introduction

Although common in vitro translation systems are well suited to study a variety of structural proteins and enzymes, they do not allow to fully exploit the potential of in vitro transcription and mutagenesis for the functional characterization of ion channels and transporters. Since 1982, when they where first used to study ligand-activated ionic channels *(1)*, *Xenopus laevis* oocytes have instead proved to be a convenient and reliable expression system for the characterization of these and other classes of membrane proteins. The popularity of this system is because of several reasons. With respect to somatic cells, *Xenopus* oocytes are easier to manipulate and require a less sophisticated experimental setup. They allow high levels of protein expression and different proteins can be easily coexpressed by simply coinjecting their respective cRNAs. In addition, because *Xenopus* oocytes express only a few types of endogenous channels, transporters and receptors, they are a nearly ideal system for the identification and characterization of exogenous proteins belonging to these classes *(2)*.

Different electrophysiological techniques, (or combinations of electrophysiological and nonelectrophysiological techniques) can be used in conjunction with the *Xenopus* oocyte expression system: patch clamp *(3)*, microinjection and two electrodes voltage clamp (TEVC) *(4)*, uptake of radiolabeled molecules and TEVC *(5)*, or fluorescence and TEVC *(6)*. Here, we will show how TEVC can be applied to the characterization of a transporter expressed in microinjected *Xenopus* oocytes.

In TEVC, the voltage across the plasma membrane is kept constant (voltage clamp) and at the same time the ionic current that permeates this membrane is measured with the help of a voltage clamp amplifier. The use of this technique is therefore restricted to the characterization of those proteins that are able to drive a net flux of charges across the plasma membrane. Because they drive ions in precise directions (downward their electrochemical gradients) relatively selective ion channels can normally be studied using the TEVC technique. The situation is more complex in the case of transporters, because some of them can simultaneously transport two or three different types of ions that may have opposite charges and that can also be transported against their electrochemical gradient. As a consequence the total charge movement across the membrane per transport cycle could be equal to zero. In this case the technique would be inadequate.

The functional characterization of heterologous transporters expressed in *Xenopus* oocytes is often achieved by using in vitro mutagenesis to introduce single or multi point mutations in the cDNA coding for the protein(s) under investigation. After in vitro transcription and microinjection of the resulting cRNAs, the effects of the introduced mutations on the characteristics of the membrane ionic current driven by the expressed protein are monitored.

As a specific example, we will describe an experiment in which a type 1 rat GABA transporter (rGAT1) is characterized after microinjection of its cRNA into *Xenopus* oocytes. rGAT1 is a $Na^+$ $Cl^-$ -dependent neurotransmitter cotransporter, which transfers 1 GABA, 2 $Na^+$, and 1 $Cl^-$ from the extracellular to the intracellular side of the plasma membrane *(7)*. At each transport cycle *per* transport protein, a positive charge enters the cell and this is monitored by recording the whole membrane current.

Although the experimental protocol described in **Subheading 3.5.** is specific for rGAT1, the general experimental setup and many of the other procedures described next would be adequate for the study of any other electrogenic transporter or ionic channel.

We will also briefly describe a recent technological development, consisting in the automation of the TEVC technique applied to *Xenopus* oocytes for drug discovery purposes.

## 2. Materials

### 2.1. Media and Solutions

1. Microelectrodes filling solution (3 *M* KCl).
2. ND96 (*see* Chapter 6).
3. Agar 3% in 3 *M* KCl.
4. Various extracellular solutions as required by individual experimental protocols (*see* **Subheading 3.4.2.**).

## 2.2. Special Items

1. Borosilicate glass capillaries (1.3 mm OD, 2.0 mm OD; Harward Apparatus Ltd., Edenbridge, Kent, UK; Garner Glass Company, Claremont, CA; World Precision Instruments, Sarasota, FL).
2. Microelectrode storage box (World Precision Instruments).
3. Small diameter 0.22-$\mu$m syringe filters.
4. Microelectrodes filling needle (Microfil™; World Precision Instruments).
5. Silver wire (0.25 and 0.37 mm OD; World Precision Instruments).
6. Ag/AgCl pellets (size dependent on the recording chamber specifications; World Precision Instruments.
7. Agar bridges (filled with agar 3% in KCl 3M; *see* **Subheading 3.2.**).
8. Plastic/silicon tubes (various sizes).

## 2.3. Equipment

1. *X. laevis* oocytes maintenance and microinjection facilities (*see* Chapter 6).
2. Solid table or vibration free table (Technical Manufacturing Corporation, Peabody, MA; World Precision Instruments)
3. Faraday cage.
4. Stereo dissection microscope (Leica, Bensheim, Germany; Zeiss, Oberkochen, Germany).
5. Recording chamber (Warner Instruments, Hamden, CT).
6. Two micromanipulators, left and right handed (Narishige, East Meadow, NY; Marzhauser Wetzlar, Wetzlar-Steindorf, Germany; Leica; EXFO Burleigh, Victor, NY; World Precision Instruments).
7. Solution exchange system (perfusion system) (Bio-Logic, Claix, France; Warner Instruments).
8. Vacuum pump.
9. Voltage clamp amplifier (Warner Instruments; Axon Instruments, Union City, CA).
10. Computer (PC or MacIntosh).
11. AD/DA converter (digitizer) (Axon Instruments; HEKA, Lambrecht, Germany, EU).
12. Electrophysiology software (Axon Instruments; HEKA; University of Strathclyde, Glasgow, UK).
13. Plot design and statistical software (Origin Lab, Northampton, MA; Sigma Plot, Point Richmond, CA).

14. Microelectrodes puller (David Kopf Instruments, Tujunga, CA; Narishige; Sutter Instruments, Novato).
15. Rack mount.

## 2.4. Experimental Setup

The reader should refer to available manuals *(8,9)* and visit an electrophysiology lab to learn the basics of how to set up an electrophysiology unit.

For the equipment, many different solutions are currently available (*see* web links). In our lab we have two slightly different setups that mainly differ in the type of voltage clamp amplifier and in the characteristics of the recording chamber. *See* **Fig. 3** for the general schematic diagram of the connections. The two systems are almost equal in terms of performance and flexibility. One of them is shown in **Fig. 1** and includes: a stereo dissection microscope wild M3B (Leica; **Fig. 2**), a Gene clamp 500B amplifier, a Digidata 1200 digitizer with pCLAMP software (Axon Instruments), a RC-1Z recording chamber (Warner Instruments; **Fig. 2**), two MM-33 micromanipulators with tilting device, left and right handed (Marzhauser Wetzlar; **Fig. 2**), a vibration-free table (Technical Manufacturing Corporation; **Fig. 1**), a TDS420 oscilloscope (Tektronix Inc., Beaverton, OR; **Fig. 1**), a 16-ways pinch valve perfusion system (homemade; **Fig. 2**), a faraday cage (homemade; **Fig. 1**), a vacuum pump (modified aquarium pump).

## 3. Methods

### 3.1. Setting Up a TEVC Experiment

A TEVC experiment essentially consists in the recording of the whole oocyte membrane current while applying membrane voltage changes and/or exchanging the extracellular solution (oocyte well solution) *(10–15)*. To accomplish this, the oocyte is placed in a recording environment (the recording chamber), which allows free access to the microelectrodes and to the ground electrodes and allows the extracellular solution to be changed by means of the perfusion system, without interfering with the recording. First you have to set up (1) the voltage clamp protocols and

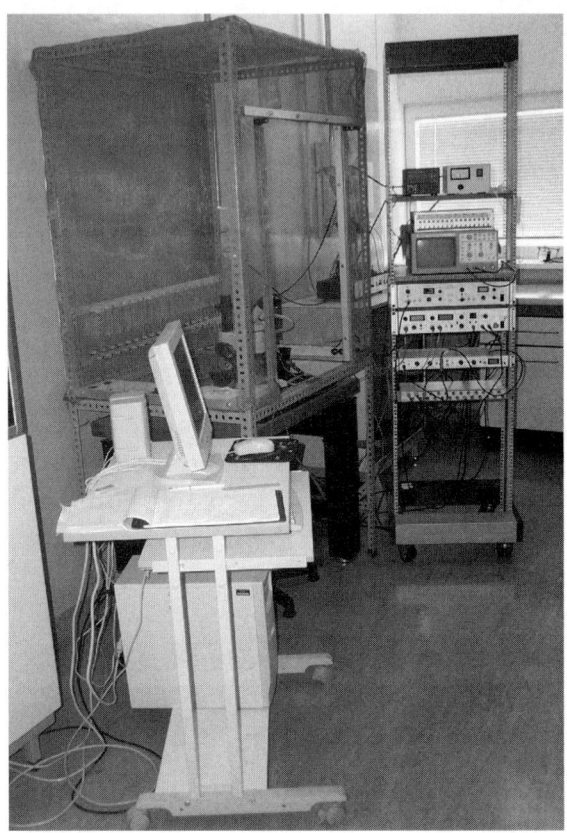

Fig. 1. Experimental setup. Front: personal computer. Middle: the vibration free table and the Faraday cage surrounding the microscope, the micromanipulators and the perfusion system. Rear: the rack mount with (front top to bottom) power suppliers, perfusion controls, oscilloscope, voltage clamp amplifier, temperature controller, electronic filter, and digitizer.

the current recording and display parameters using the pClamp software, (2) the automated perfusion system (if your system has one). All these parameters can then be modified when required during the actual experiment. The characteristics of the voltage protocols and the composition of the extracellular solutions will be specific for each experiment and will depend on the type of protein under study.

In the next two steps (**Subheadings 3.2.** and **3.3.**) we will describe how to ensure the electric connectivity between the two sides of the oocyte membrane and the voltage clamp amplifier. Particular care should

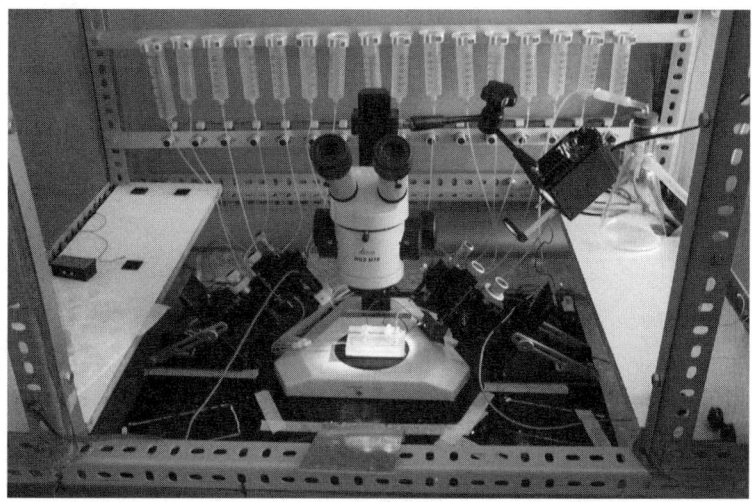

Fig. 2. Inside the cage. Front: the Farady cage aperture. Middle: the microscope with the recording chamber already positioned and flanked by the two micromanipulators. Rear: 30-mL syringes vertically aligned with the electrovalves of the perfusion system.

be taken in performing these critical steps. A single defective component in the electrical chain connection will impair the recording or introduce undesired artifacts.

Some of the next steps should be accomplished in advance, hours or days (*see* **Subheading 4.** for details), in order to save time for the recordings (the experiment) as oocytes have a relatively short life after microinjection and you must be able to optimize their use.

## 3.2. Recording Chamber and Agar Bridge Preparation

The simplest way to guarantee an electrical connectivity between a solution and an electric wire is to use an Ag-AgCl electrode dipped in the solution. In this case $Cl^-$ must be present in the solution as the charge transporting ion between the solution and the Ag-AgCl electrode. However, some experimental protocols require to work in absence of extracellular chloride. In this case two agar bridges must be used to ensure electrical connectivity between the two bath electrodes and the bath solution and the connectivity scheme will be: voltage clamp amplifier–Ag-AgCl pellet–KCl 3M solution (in the agar bridge wells)–agar

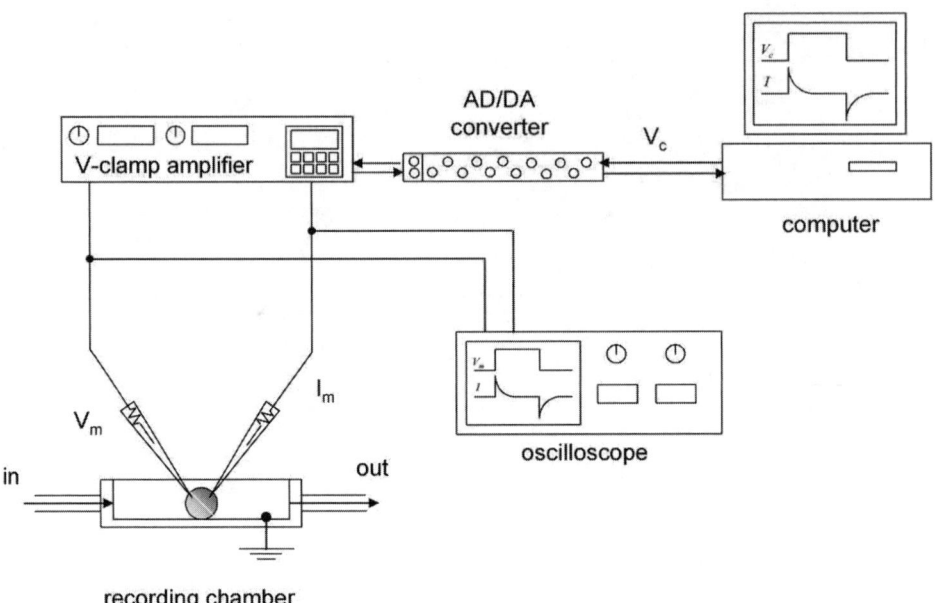

Fig. 3. Schematic diagram of connections between the instruments of the TEVC setup.

bridge–bath solution (in the oocyte well/outlet well) (**Fig. 4**) (*see* **Note 1 and Subheading 3.4.1.**).

1. Fix the two Ag-AgCl pellets to the recording chamber's agar bridge wells (*see* **Note 2**).
2. On a Bunsen burner, curve a short piece of glass capillary (2 mm OD) to obtain an approximate U shape (*see* **Note 3**).
3. Insert a piece of silver wire (0.37 mm OD) inside the glass capillary and fix it on both sides by bending it by 180° around the capillary holes (*see* **Note 4**).
4. Prepare the Agar-KCl gel solution (*see* **Note 5**).
5. Fill the curved glass capillary with the agar-KCl gel solution (*see* **Note 6**).

### *3.3. Microelectrodes Preparation*

1. Cut borosilicate glass capillaries to the desired length (*see* **Note 7**).
2. Pull the microelectrodes following the puller reference manual (*see* **Note 8**).
3. Store the pulled microelectrodes in the storage box (*see* **Note 9**).

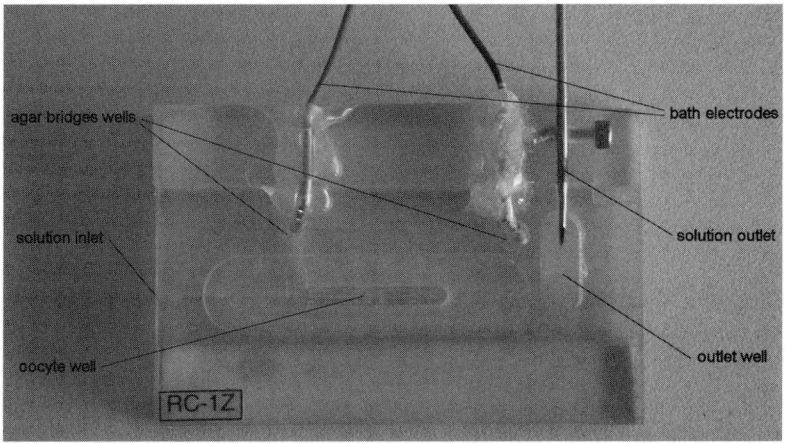

Fig. 4. Recording chamber.

## 3.4. The TEVC Experiment

Oocytes are prepared and microinjected (*see* Chapter 6) with 12.5 ng/ oocyte of mRNA encoding rGAT1 *(16)*. Some of the oocytes will not be injected at all or alternatively will be injected with water and kept separately as control. Recordings are performed 2–3 d after microinjection to allow the protein to be expressed and inserted in the plasma membrane (*see* **Note 10**).

### 3.4.1. Checking the Control Oocytes

When starting an experiment (or if the batch of oocytes is changed during the experiment) we recommend to check for the general status of noninjected/water-injected oocytes first, as described in the following steps, and only later to proceed with mRNA injected oocytes (**Subheading 3.4.2.**).

1. Fill the tube no. 1 of the perfusion system with the ND96 solution (*see* **Note 11**).
2. Fill the agar bridge wells with 3 *M* KCl solution and the recording well with the extracellular solution ND96 (*see* Chapter 6).
3. Connect the two bath electrodes (**Fig. 4**) to the "bath reference/headstage" provided with the amplifier (*see* the voltage clamp amplifier reference manual).

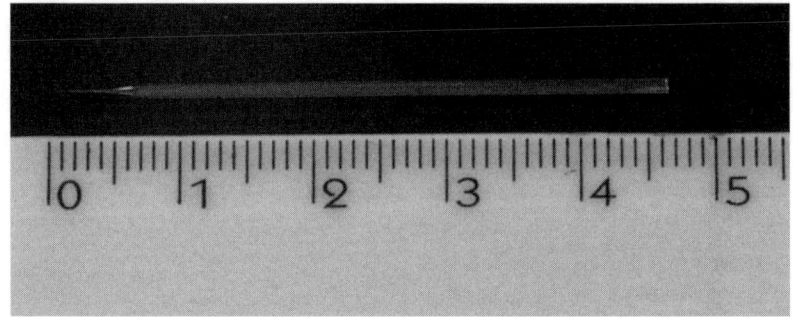

Fig. 5. Pulled glass microelectrode appearance (centimeters).

4. Mount the two agar bridges so they establish electrical connection between the oocyte and the outlet wells and their respective agar bridge wells (**Fig. 4**) (*see* **Note 12**).
5. Fill the two microelectrodes (**Fig. 5**) with 3 *M* KCl solution (*see* **Note 13**).
6. Mount the two microelectrodes on their holders and fit the holders on the amplifier headstages (*see* **Note 14**).
7. Place an oocyte on the bottom of the oocyte well in a convenient position using the modified Pasteur pipet (*see* Chapter 6) (*see* **Note 15**).
8. Sink the two microelectrodes in the bath solution maneuvering the micro-manipulators.
9. Break the microelectrodes tips (gently crashing them on the bottom/sides of the recording chambers) to obtain about 1 and 4 MOhm resistance for the current and for the voltage microelectrodes respectively (*see* **Note 16**).
10. Set to zero the tip potential of both microelectrodes (following the amplifier reference manual).
11. Gently move the micromanipulator controls to push the electrodes in close proximity of the oocyte membrane (*see* the microinjection procedure in Chapter 6).
12. Start to impale the oocyte, first with the voltage microelectrode and then with the current one (*see* **Note 17**).
13. The two microelectrodes should measure approximately the same voltage across the oocyte membrane (*see* **Note 18**).
14. Position the tip of the perfusion system to align it with the oocyte.
15. Switch on the voltage clamp circuit.
16. Set the desired holding potential (clamp potential) and apply a membrane test pulse (*see* **Note 19**).

17. Set the right speed of the voltage clamp circuit (*see* **Note 20**).
18. Check the resting (leakage) current (*see* **Note 21**).

**Steps 7–16** are generally described in details in the voltage clamp amplifier reference manual, as the current/voltage meters, knobs, and buttons layout is somehow different between amplifiers brands. *See also* **refs.** *8* and *9*.

## 3.4.2. Recording of the Transport-Associated Current in rGAT1

Fill the tubes no. 1 and 2 of the perfusion system respectively with the ND96 solution (solution1) and with the ND96 added by GABA at the final concentration of 100 $\mu M$ (solution 2).

1. Start the pCLAMP software and load a protocol suitable for relatively long recordings (*see* **Note 22**).
2. Go through **steps 2–18** of **Subheading 3.4.**
3. Clamp the oocyte membrane voltage at –40 mV.
4. Switch on the vacuum pump (*see* **Note 23**).
5. Perfuse the oocyte with solution 1 (*see* **Note 24**).
6. Start the recording.
7. Wait the baseline current to remain stable (relatively flat line) (*see* **Note 25**).
8. Start the perfusion with solution 2 and stop perfusing with solution 1 (at the same time) and wait until the current reaches a plateau level.
9. Stop the perfusion with solution 2 and start perfusing with solution 1 (at the same time).
10. Wait for the current to come back close to the resting value.
11. Stop the recording (*see* **Note 26**) (**Fig. 6**).

The plotted data in **Fig. 6** show the activation of an inward membrane current following the extracellular application of GABA at the concentration of 100 $\mu M$. This behavior is consistent with the functional expression of rGAT1 GABA cotransporter, which couples a positive charge ($Na^+$) movement *per* GABA molecule transferred inside the cytoplasm.

Two current spikes, coincident with the changes of the extracellular solution, are clearly visible in the current trace. These are artifacts owing to the opening of electrovalves and can be removed for aesthetic reasons only.

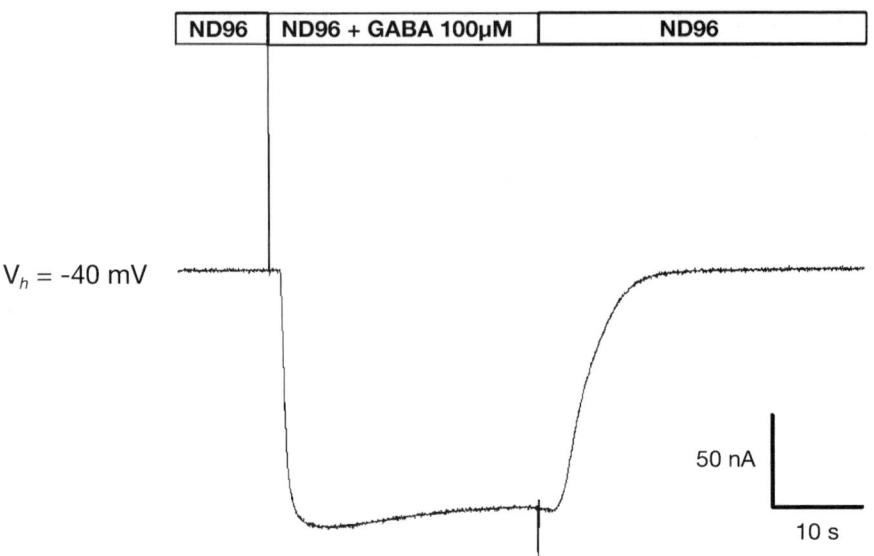

Fig. 6. Transport associated current. Whole oocyte current trace before, during, and after the perfusion of GABA at the concentration of 100 µ$M$, holding (clamp) potential $V_h = -40$ mV. The bar on top represents the timing of the perfusion action.

## 3.5. Oocyte-Based High-Throughput Screening Systems

### 3.5.1. TEVC Recordings From Xenopus Oocytes: Application in Ion-Channel Drug Discovery

The search for new drugs generally starts with the set up of an automated high-throughput screen (HTS) (*see* **Note 27**). The voltage-clamp technique in both variants of the TEVC applied on the *Xenopus* oocytes (TEVC) and of the patch clamp, allows for the membrane potential to be clamped at specific voltages with a precise control of the kinetic responses and intrinsic properties of ion channels, such as the time constants for channel opening, closing, and inactivation. Pharmacological modulation of ionic currents can be studied under well-defined conditions. This allows one to monitor the state-dependent behavior of ion channel blockers and, ultimately, provides information on its mechanism of action. However, the limited throughput of these methods constitutes a serious bottleneck in the drug discovery process and this motivated various efforts toward their automation *(17,18)* (*see* **Note 28**).

Based on the robustness and extreme versatility of the *Xenopus* oocyte expression system and considering the high information content provided by electrophysiological approaches to the pharmacological characterization of ion channels, it appears reasonable to consider a partial automation of the oocyte handling procedures to exploit this valuable method as a screening tool to be used for drug discovery by the pharmaceutical industry.

### 3.5.2. An Oocyte-Based Workstation For Screening Ion Channel Targets: The Roboocyte™

Among the various electrophysiological techniques, TEVC recordings in *Xenopus* oocytes seems particularly suitable for the implementation of automated measurement systems. Here, we describe a workstation that was expressly developed for this purpose. The Roboocyte produced by Multi Channel Systems MCS GmbH (Reutlingen, Germany) is the first (and the only one currently available) instrument that automatically performs both cDNA and mRNA injections and subsequent TEVC recording from *Xenopus* oocytes in a standard 96-well microtiter plate format *(19)*. The workstation includes the Roboocyte robot, a gravity-based 16-pinch valve perfusion system, the Roboocyte data acquisition and analysis software package, and other accessories like ready-to-use disposables TEVC probes and standardized injection pipets.

In the following section the general operating modes of the Roboocyte will be briefly described. Because most of automated operations are preset in the device, the user will consult the Roboocyte Manual for more detailed descriptions.

The voltage and compound application protocols are edited by the user as instruction lines in script protocols according to the specific target and to the specific experimental design. The script language rules are described in a Script Manual provided with the device.

### 3.5.3. General Methods and Operations

Injection, recording, and cultivation of *Xenopus* oocytes is performed using disposable standard 96 V-shape well plates. After preparation and selection of the oocytes, as described in the Chapter 6, these are plated into the cone-shaped wells (one oocyte per well) where they quickly settle and adhere to the well bottom after a few hours. The oocytes can be kept for several days in the plate. Plates can be transferred from the incubator to the Roboocyte for the injection and recording operations, and then back again to the incubator.

The well plate carrier of the Roboocyte, powered by linear motors, hovers smoothly and noise free on a cushion of compressed air above the magnetic $x/y$ table. The carrier operates at a resolution of 20 μm. The two vertically moving z-arms, one holding the injection needle, one the TEVC probe, are designed specifically to provide both high speed and precision. They move at a resolution of 33 μm; position and speed are computer-controlled. A quick adjustment process guarantees that the oocytes are injected/or impaled for recording precisely.

In the injection mode the cDNA/mRNA to be injected is filled into the injection pipet and then pushed into either the nucleus or the cytoplasm, respectively, by compressed air. Prepulled borosilicate glass micropipets are provided by Multi Channel Systems, but custom injection needles may also be used. The basic script protocols for the injection can be modified by the user according to the specific needs.

The ClampAmp is a specifically designed digital TEVC amplifier and is completely automated. Ready to use TEVC probes are also available.

The experimental protocols driving the recording and the compound testing are edited in script protocols by the customer. The commands in the scripts drive the different steps of the experiments (membrane potential and leakage checks, application of voltages and compounds) and automatically consider if acceptance criteria are established during a single experiment to proceed to next steps of the protocol or to move to another oocyte.

The integrated pinch valve system is a gravity-driven perfusion system for drug receptor characterizations and quick expression tests. The on and off settings for electrovalves and external suction pump are computer (scripts) controlled.

In conclusion the Roboocyte allows the automation of different steps of the experimental procedure for current recording and compound testing, improving 10-fold the capability for functional-based drug screening with respect to a classical TEVC setup.

## 4. Notes

1. The following operations should be accomplished well in advance before the actual experiment is performed.
2. The two Ag-AgCl pellets are positioned on the bottom of the agar bridges wells and fixed in position with a drop of glue. The electrical connection between the two pellets and the voltage clamp amplifier is obtained soldering the wire emerging from the Ag-AgCl pellets to the two copper wires that will be then connected to the ground headstage.
3. The bridge length should be enough to span the distance between the agar bridges wells and the oocyte or outlet wells.
4. This operation will increase the conductivity of the bridge.
5. The Agar-KCl gel solution is prepared solubilizing on a heated stirrer 3% agar in a 3 *M* KCl solution.
6. Before the Agar-KCl gel solution gets too thick (while cooling) fill the curved glass capillary with the solution avoiding the formation of air bubbles. Do so by sucking the KCl/agar solution inside the capillary applying a negative pressure, using a syringe fitted with a silicon tubing. The agar bridges must be stored in KCl 3 *M* solution when not in use. They can last for months under these storing conditions.
7. Different types of glass and capillaries can be used. Standard borosilicate for electrophysiology is suitable for most of the TEVC experiments and glass with specific compositions are available for critical protocols (mostly related to path clamp) *(8,9)*. Capillaries from 1.0 to 2.0 mm OD can be used, but lower diameters (1 to 1.3 mm OD) are preferred to obtain lower diameter tips and reduce membrane damage during oocyte impalement. Check also the specifications of the microelectrode holder provided with the voltage clamp amplifier, as you can generally choose between different sizes when ordering. To facilitate the filling of the electrode tip and avoid the formation of air bubbles, we suggest using commercially available capillaries with an inner thin glass filament.
8. No particular attention should be paid to microelectrode resistance at this point, as the correct resistance value has to be set just before oocyte impalement (*see* **Subheading 3.4.1.**). Pull electrodes with a resistance around 100 MOhms or greater when filled with the electrode solution (*see* **Fig. 5** for a common electrode tip shape). With practice, you will not have to check the resistance of each microelectrode. You will understand when a microelectrode is good simply by looking at its shape.

9. The microelectrodes are generally pulled the same day of the experiment to avoid contact with dust particles that can obstruct the electrode tip. For the same reason, they are kept in a sealed storage box until used.

10. The period between injection and appearance of maximal expression will vary depending on the characteristics of the protein. There are no rules to anticipate how long this period could be and this must be established experimentally performing daily measurements of protein expression at the plasma membrane, in our case measuring the mean transmembrane current.

11. The perfusion solutions should be prepared in advance (days) and stored properly, generally at 4°C. They will be loaded in the perfusion syringes at the beginning of the experiment. Prewarm the solutions before filling the perfusion tubes to avoid the formation of air bubbles that could obstruct the perfusion system.

12. One of the two reference (ground) electrodes should be positioned in proximity of the oocyte (oocyte well), for electronic circuitry reasons. Read the voltage clamp amplifier reference manual to understand how to correctly connect the reference electrodes.

13. Back fill the microelectrodes (starting from the tip) to about half of their length using a heat pulled P200 plastic tip applied to a syringe or better using a Microfil nonmetallic syringe needle. To avoid dirt to obstruct the microelectrode tip apply a 0.22-μm filter between the syringe and the needle. Tap gently the microelectrode if air bubbles are present. Air bubbles must be removed because they can cause an open circuit. Fill the microelectrodes just before their use as evaporation (KCl crystallization) at the tip can cause microelectrode obstruction.

14. Before inserting the electrodes in their holders check the integrity of the AgCl coating on the silver wire (dark gray) to obtain the correct electric connectivity between the amplifier headstage and the KCl solution. Ag wires can be chlorided by sinking them in common bleach for 15–30 min. This operation should be performed periodically.

15. Most of the chambers for oocyte recordings come with holes to facilitate the correct positioning of the oocyte. If this is not the case, you should create them, otherwise the oocyte will slip away when you try to penetrate it. You can make holes at the bottom of the chamber or you can glue at the bottom a piece of plastic mesh. In any case the holes size should be about two-thirds of the mean oocyte diameter.

16. Check the resistance using the resistance check button on your amplifier (*see* the voltage clamp amplifier instruction manual). Glass microeletrodes can be used several times (with different oocytes) until they get obstructed by membrane and/or cytoplasmic material. Always check the microelec-

trode resistance before impalement of each oocyte to understand if micro-electrodes replacement is needed.

17. You have successfully penetrated the membrane (i.e., established an electrical connection with the cytoplasm) when you read a voltage change on the amplifier meters (*see* the amplifier reference manual).

18. Typically membrane voltage values range from –20 to –60 mV in ND96 solution, in healthy uninjected oocytes.

19. The membrane test pulse is a square voltage pulse of about 10 mV from the resting potential.

20. The speed of the voltage clamp circuit should be set as fast as possible, avoiding oscillations or ringing of the circuit.

21. The resting current should be as small as possible, typically –10/–50 nA in uninjected oocytes at –40 mV clamped potential. Larger currents are an indication of a damaged oocyte membrane owing to a bad impalement or to a nonhealthy oocyte (with damaged plasma membrane). In injected oocytes the resting current could have higher values in relation to the characteristics of the ionic channel/transporter you are exogenously expressing. It is therefore good practice to check first also few noninjected oocytes (or water-injected oocytes) from the same batch.

22. The experimental voltage protocols must be constructed in advance with the pCLAMP software (*see* pCLAMP reference manual). In this experiment a constant voltage must be applied and a current "gap free" recording (without time limits) should be performed. Choose the appropriate sampling frequency in relation to the frequency of your signal (*5*), set to 40 Hz in this case. Switch the low pass filter on the current output of the amplifier to 20 Hz.

23. This operation prevents recording chamber solution spill over while perfusing the oocyte

24. You should check that the flux of solution does not damage the oocyte and does not induce variations of the baseline current (leakage current). Be aware of changes in solution levels in the recording chambers as they can have effect on the baseline current amplitude and/or electric noise levels.

25. Quite commonly the baseline undergoes initial amplitude variations of small entity, related to improvement or deterioration of the seals between electrodes and the plasma membrane over time. This value should remain as constant as possible.

26. You are now ready to analyze/plot your data as required.

27. Abnormal ion channel functions or reduced ion channel expression have been linked to a number of diseases, including cardiac arrhythmia, hyper-

tension, anxiety, epilepsy, pain, neuroprotection, and diabetes. Novel ion channel modulator drugs for these therapeutic areas have yet to be developed, leading to a growing demand for high throughput ion-channel functional assays. Chemioluminescence or fluorescence assays are mostly employed to correlate changes in the concentration of a particular ion or in membrane potential with light intensity and work essentially well for detecting changes in intracellular calcium concentrations. However, with the currently available dyes not all types of ion channels can be investigated *(17)*. The most important handicap is that the voltage across the cell membrane cannot be adequately controlled. In addition, most HTS assays for ion channels do not permit a clear differentiation of a compound's effect on different subtypes of ion channels and do not reveal which specific biophysical properties of an ion channel are modulated. To answer these questions a secondary and safety screening must be set up using electrophysiological methods that provide the most direct and sensitive measurement of ions flowing through ion channels in individual cells.

28. Traditionally these techniques have not been used for HTS of ion channel targets because electrophysiology is labor-intensive and a low throughput process, not well fitting with the industry research program timelines. Ion channel testing has generally been performed by highly skilled technicians using one oocyte/or cell at time. Equipment setup and maintenance also requires a high level of expertise. Analysis and organization of the results could be laborious. Therefore, TEVC and the patch clamp on *X. laevis* oocytes have been largely restricted to laboratories already skilled in the practice of electrophysiology for basic research.

29. Useful Web links:
Technical Manufacturing Corporation
http://www.techmfg.com/company/copyright_info.html
World Precision Instruments
http://www.wpiinc.com/
Warner Instruments
http://www.warnerinstrument.com/
Narishige
http://www.narishige.co.jp/
Marzhauser Wetzlar
http://www.marzhauser.com/index.html
Leica
http://www.leica-microsystems.com/website/lms.nsf
EXFO Burleigh
http://www.exfo.com/en/index.asp

Bio-Logic
http://www.bio-logic.info/index.html
Axon Instruments
http://www.axon.com/index.html
HEKA
http://www.heka.com/
CAIRN (Faversham, Kent, UK)
http://www.cairnweb.com/menus/menustub_main.html
Origin Lab
http://www.originlab.com/
Sigma Plot
http://www.systat.com/products/SigmaPlot/
David Kopf Instruments
http://www.kopfinstruments.com/
Sutter Instruments
http://www.sutter.com/
Harward Apparatus LTD
http://www.clark.mcmail.com/
Garner Glass Company
http://www.garnerglass.com/
Tektronix Inc.
http://www.tek.com/
Strathclyde electrophysiology software (Dr. John Dempster, Department of Physiology and Pharmacology, University of Strathclyde)
http://innovol.sibs.strath.ac.uk/physpharm/ses.shtml
Roboocyte (Multi Channel System MCS GmbH, Reutlingen)
http://www.multichannelsystems.com/

# References

1. Miledi, R., Parker, I., and Sumikawa, K. (1982) Properties of acetylcholine receptors translated by cat muscle mRNA in *Xenopus* oocytes. *EMBO J.* **1**, 1307–1312.
2. Weber, W. M. (1999) Ion currents of *Xenopus laevis* oocytes: state of the art. *Biochim. Biophys. Acta.* **1421**, 213–233.
3. Lu, C. and Hilgemann, D. W. (1999) GAT1 (GABA:Na+:Cl-) cotransport function. Steady state studies in giant *Xenopus* oocyte membrane patches. *J. Gen. Physiol.* **114**, 429–444.

4. Centinaio, E., Bossi, E., and Peres, A. (1997) Properties of the $Ca^{2+}$-activated $Cl^-$ current of Xenopus oocytes. *Cell. Mol. Life Sci.* **53,** 604–610.

5. Bossi, E., Vincenti, S., Sacchi, V. F., and Peres, A. (2000) Simultaneous measurements of ionic currents and leucine uptake at the amino acid cotransporter KAAT1 expressed in *Xenopus laevis* oocytes *Biochim. Biophys. Acta.* **1495,** 34–39.

6. Larsson, H. P., Tzingounis, A. V., Koch, H. P., and Kavanaugh, M. P. (2004) Fluorometric measurements of conformational changes in glutamate transporters. *Proc. Natl. Acad. Sci. USA* **101,** 3951–3956.

7. Masson, J., Sagne, C., Hamon, M., and El Mestikawy, S. (1999) Neurotransmitter transporters in the central nervous system. *Pharmacol. Rev.* **51,** 439–464.

8. Sakmann, B. and Neher, E. Single channel recording 2^nd edition, Plenum Press NY

9. Various authors. The Axon Guide (electrophysiology and biophysics laboratory techniques), downloadable from the Axon Instrument web site http://www.axon.com/mr_Axon_Guide.html

10. Cole, K. S. and Curtis, H. J. (1941) Membrane potential of the squid giant axon during current flow. *J. Physiol.* **100,** 551–563.

11. Hodgkin, A. L., Huxley, A. F., and Katz, B. (1952) Measurements of current-voltage relations in the membrane of the giant axon of Loligo. *J. Physiol.* **116,** 424–448.

12. Hodgkin, A. L. and Huxley, A. F. (1952) A quantitative description of membrane current and its application to conduction and excitation in nerve. *J. Physiol.* **117,** 500–544.

13. Marmont, G. (1949) Studies on the axon membrane: I. A new method. *J. Cell. Comp. Physiol.* **34,** 351–382.

14. Cole, K. S. (1949) Dynamic electrical characteristics of the squid axon membrane. *Arch. Sci. Physiol.* **3,** 253–258.

15. Hamill, O. P., Marty, A., Neher, E., Sakmann, B., and Sigworth, F. J. (1981) Improved patch-clamp techniques for high resolution current recording from cells and cell-free membrane patches. *Pflug. Arch.* **391,** 85–100

16. Forlani, G., Bossi, E., Perego, C., Giovannardi, S., and Peres, A. (2001) Three kinds of currents in the canine betaine-GABA transporter BGT-1 expressed in *Xenopus laevis* oocytes. *Biochim. Biophys. Acta.* **1538,** 172–180.

17. Xu, J., Wang, X., Ensign, B., et al. (2001) Ion channel assay technologies: quo vadis? *Drug Discovery Today* **6,** 1278–1287.

18. Wang, X. and Li, M. (2003) Automated electrophysiology: high through-put of art. *Assay and Drug Development Technologies* **1,** 695–708.

19. Schnizler, K., Küster, M., Methfessel, C., and Fejtl, M. (2003) The roboocyte: automated cDNA/mRNA injection and subsequent TEV recording on Xenopus oocytes in 96-well microtiter plates. *Receptors and Channels* **9,** 41–48.

# 13

## Production of Protein for Nuclear Magnetic Resonance Study Using the Wheat Germ Cell-Free System

**Toshiyuki Kohno and Yaeta Endo**

### Summary

Nuclear magnetic resonance (NMR) methods have been developed to determine the three-dimensional structures of proteins, to estimate protein folding, and to discover high-affinity ligands for proteins. However, one of the difficulties encountered in the application of such NMR methods to proteins is that we should obtain milligram quantities of $^{15}N$ and/or $^{13}C$-labeled pure proteins of interest.

Here, we describe the method to produce proteins for NMR experiments using the improved wheat germ cell-free system, which exhibits several attractive features for high-throughput NMR study of proteins.

**Key Words:** Wheat germ; cell-free; protein synthesis; NMR; HSQC; protein folding; transaminase; glutamine synthase; selective labeling.

## 1. Introduction

Nuclear magnetic resonance (NMR) methods have been developed to determine the three-dimensional structures of proteins *(1–3)*, to estimate protein foldings *(1)*, and to discover high-affinity ligands for proteins *(4)*. However, one of the limitations of NMR methods to structural studies of proteins is the requirement of milligram quantities of $^{15}N$ and/or $^{13}C$-

From: *Methods in Molecular Biology,* vol. 375,
*In Vitro Transcription and Translation Protocols, Second Edition*
Edited by: G. Grandi © Humana Press Inc., Totowa, NJ

labeled highly purified proteins. To prepare stable-isotope labeled proteins, *Escherichia coli* over-expression systems have been widely used because they often provide large quantities of the protein of interest, leading to a low cost for labeling. However, *E. coli* over-expression systems have some drawbacks: (1) proteins that interfere with the physiology are hard to express, (2) expressed proteins require extensive purification before NMR measurements to eliminate contaminants from host cells that are inevitably colabeled with stable-isotope, (3) eukaryotic proteins often do not assume correct folding. Recently, a novel wheat germ cell-free protein synthesis system has been developed *(5,6)*. The advantage of this system is its ability to produce eukaryotic proteins in their native conformation. In addition, the system also exhibits several attractive features for preparing NMR samples: (1) it is robust, because the reaction mixture is stable over long periods as it contains low amounts of degradation enzymes such as proteases or ribonucleases *(6)*, (2) it can produce milligram quantities of the protein of interest per milliliter of reaction volume *(6,7)*, (3) when it is used for stable-isotope labeling, only the protein of interest can be labeled, therefore the synthesized protein may be analyzed by hetero-nuclear NMR measurements without purification *(7)*, (4) when it is used with some inhibitors for amino acid metabolic enzymes, the amino acid selective labeling may be performed *(8)*.

Here, we describe the isotope-labeling of the yeast protein ubiquitin by the wheat germ in vitro expression system and the use of the non-purified protein in NMR studies.

## 2. Materials

1. Yeast ubiquitin cDNA.
2. *E. coli* strain JM109.
3. Restriction enzymes, T4 DNA polymerase, and T4 DNA ligase.
4. LB media and ampicillin.
5. pEU3b vector (*see* **Fig. 1** and **Note 1**) *(6)*.
6. 1 *M* HEPES-KOH, pH 7.8, magnesium acetate, spermidine trihydrochloride, dithiothreitol (DTT), and RNase-free water.
7. ATP, UTP, CTP, and GTP.
8. SP6 RNA polymerase and RNase inhibitor (Takara, Kyoto, Japan or Promega, Madison, WI).

9. RiboGreen RNA Quantitation Kit (Molecular Probes, Eugene, OR).
10. Potassium acetate and creatine phosphate.
11. Amino acids and $^{15}$N-labeled amino acids (Cambridge Isotope Laboratories, Andover, MA).
12. Wheat germ extract (Cell Free Science or ZOEGENE Corporation, Yokohama, Japan).
13. Creatine kinase (Roche, Indianapolis, IN).
14. Wheat germ transfer RNA (Sigma-Aldrich, St. Louis, MO) (*see* **Note 2**).
15. NMR buffer: 50 m*M* sodium phosphate, pH 6.5, 0.1 *M* NaCl, and 4, 4-dimethyl-4-silapentane-L-sulfonic acid (DSS).

## 3. Methods

### 3.1. Plasmid Template for Wheat Germ Cell-Free System

#### 3.1.1. Cloning

DNA manipulations were performed by standard recombinant DNA methods *(9)*. The pEU3b plasmid was digested with *Spe*I, blunted with T4 polynucleotide kinase and then digested with *Sal*I. The pET-24a/ubiquitin *(10)* carrying the ubiquitin gene was digested with *Nde*I, blunted with T4 polynucleotide kinase, and then digested with *Sal*I. The digested pEU3b plasmid and the coding ubiquitin region were ligated with T4 DNA ligase to produce pEU3b/ubiquitin. After transformation of *E. coli* JM109 cells, the plasmid DNA from positive clones was isolated and checked for the presence and the correct orientation of the insert using restriction enzyme and DNA sequencing analyses.

#### 3.1.2. Plasmid Preparation

The JM109 cells harboring the plasmid pEU3b/ubiquitin were cultured at 37°C overnight in 100 mL of LB medium containing 50 μg/mL ampicillin. The cells were harvested and the plasmid was extracted according to the standard alkaline lysis with SDS method *(9)*. The extracted plasmid was further purified according to equilibrium centrifugation in CsCl-ethidium bromide gradients method *(9)*. Alternatively, the plasmid pEU3b/ubiquitin may be purified with ion-exchange resin commercially available (Wizard Plus SV, Promega) *(9)* (*see* **Note 3**). Finally, the plasmid was dissolved in RNase-free water. The final concentration of the plasmid was adjusted at 1.0 μg/μL.

## 3.2. mRNA Transcription for Wheat Germ Cell-Free system

### 3.2.1. mRNA Transcription

1. Prepare 400 μL of transcription solution for the yeast ubiquitin mRNA by mixing the ingredients listed in **Tables 1** and **2**.
2. Incubate the transcription solution for 5 h at 40°C. White precipitant can be seen as a result of magnesium pyrophosphate after the reaction (*see* **Note 4**).
3. Centrifuge the transcription solution (20,000*g*; 5 min) at 4°C and remove the white pellet.
4. Put the supernatant on ice, and add 110 μL of 7.5 *M* ammonium acetate and 1.0 mL of 100% ethanol (*see* **Note 5**). Put the tube on ice for 10 min. Centrifuge the solution (20,000*g*; 20 min) at 4°C and discard the supernatant. Rinse the pellet with about 300 μL of ice cold 70% ethanol, then centrifuge (20,000*g*; 1 min) again. Remove the supernatant and dry the pellet. Resuspend the pellet with 200 μL of RNase-free water.
5. The synthesized mRNA can be stored at −80°C for several weeks.

### 3.2.2. Quantitation of Synthesized mRNA

1. Dilute the RiboGreen Reagent (Molecular Probes) 200-fold. For example, to prepare enough working solution to assay five samples in 200 μL volumes, add 2.5 μL RiboGreen RNA quantitation reagent to 497.5 μL 10 m*M* Tris-HCl, pH 7.5, 1 m*M* EDTA (TE).
2. Prepare a 2-μg/mL solution of ribosomal RNA standard provided in TE. Prepare a series of standard RNA at the concentration of 0–2 μg/mL. Mix the equal volumes (100 μL) of diluted RiboGreen Reagent and standard RNAs. Measure the sample fluorescence using a TD-700 fluorometer (excitation ~480 nm, emission ~520 nm) and prepare the standard curve.
3. Take 5 μL of the transcribed ubiquitin mRNA and dilute 200-fold with TE. Mix the diluted mRNA with 100 μL of RiboGreen Reagent, and measure fluorescence. Determine the concentration of the mRNA by comparing the intensity of fluorescence with the standard curve previously determined.

**Table 1**
**5X Transcription Buffer (TB)**

| Stock | Reagent | Final concentration | |
|---|---|---|---|
| 1.0 *M* | Hepes-KOH, pH 7.8 | 400 m*M* | 400 µL |
| 1.0 *M* | Magnesium acetate | 80 m*M* | 80 µL |
| 100 m*M* | Spermidine trihydrochloride | 10 m*M* | 100 µL |
| 1.0 *M* | Dithiothreitol | 50 m*M* | 50 µL |
| | RNase-free water | | 370 µL |
| | | Total | 1.0 mL |

4. If you want to verify that the mRNA is not degraded by trace contaminations of RNases, analyze the mRNA by agarose gel electrophoresis using standard protocol *(9)*. Usually, ladder bands or smear bands of approx 1–3 kb are visible when good mRNA is obtained.
5. Adjust the concentration of mRNA at about 0.5–1.0 µg/µL. The mRNA can be stored for several weeks at –80°C.

## 3.3. Protein Synthesis by Wheat Germ Cell-Free Protein Synthesis System

### 3.3.1. Preparation of Uniformly Labeled Proteins for NMR Analyses

1. Prepare 5X-translation solution described in **Table 3**. 5X-translation solution can be stored at –20°C
2. Mix 200 µL of 5X translation solution, 325 µL of wheat germ extract (OD = 150) (*see* Chapter 5 for wheat germ extract preparation), 10 µL of wheat germ tRNA (20 µg/µL), 10 µL of creatine kinase (40 µg/µL), 14 µL of RNase inhibitor (40 U/µL), and 165 µg of the transcribed yeast ubiquitin mRNA prepared in **Subheading 3.2.** Adjust the volume to 1.0 mL with RNase-free water.
3. Load the 1-mL DispoDialyzer (Spectrum Laboratories) tube with the translation solution. Dialyze the solution against about 15 mL of the dialysis buffer (*see* **Table 4**) in a sterile tube at 26°C for 2 d while stirring the dialysis buffer.
4. After 2 d, add concentrated yeast ubiquitin mRNA to the translation solution and change the dialysis buffer. Incubate at 26°C for and additional 2 d while stirring the dialysis buffer (*see* **Note 6**).

**Table 2**
**Transcription Solution**

| Stock | Reagent | Final concentration | |
|---|---|---|---|
| 5X | 5X TB | 1X | 80 μL |
| 25 m*M* each | ATP, CTP, GTP, UTP | 5.0 m*M* | 80 μL |
| 70 U/μL | RNase inhibitor | 1.0 U/μL | 6 μL |
| 50 U/μL | SP6 RNA polymerase | 1.5 U/μL | 12 μL |
| 1.0 μg/μL | pEU3b/ubiquitin | 0.1 μg/μL | 40 μL |
| | RNase free water | | 182 μL |
| | | Total | 400 μL |

**Table 3**
**5X Translation Solution**

| Stock | Reagent | Final concentration | |
|---|---|---|---|
| 1.0 *M* | Hepes-KOH (pH7.8) | 100.0 m*M* | 100 μL |
| 4.0 *M* | Potassium acetate | 384.0 m*M* | 96 μL |
| 1.0 *M* | Magnesium acetate | 8.0 m*M* | 8 μL |
| 100 m*M* | Spermidine trihydrochloride | 2.0 m*M* | 20 μL |
| 1.0 *M* | Dithiothreitol | 10.0 m*M* | 10 μL |
| 2.5 m*M* each | $^{15}$N-labeled amino acids | 1.2 m*M* | 480 μL |
| 100 m*M* | ATP | 6.0 m*M* | 60 μL |
| 20 m*M* | GTP | 1.3 m*M* | 66 μL |
| 500 m*M* | Creatine phosphate | 80.0 m*M* | 160 μL |
| | | Total | 1.0 mL |

## 3.3.2. Preparation of Selectively Labeled Proteins for NMR Analyses

1. To label $^{15}$N-Ala or $^{15}$N-Asp selectively, add 1 m*M* aminooxyacetate to both the translation solution and the dialysis buffer. All the other procedures are the same as those written in **Subheading 3.3.1.** (*see* **Note 7**).
2. To label $^{15}$N-Glu selectively, add 1 m*M* aminooxyacetate and 0.1 m*M* L-methionine sulfoximine to both the translation solution and the dialy-

**Table 4**
**Dialysis Buffer**

| Stock | Reagent | Final concentration | |
|---|---|---|---|
| 1.0 *M* | Hepes-KOH (pH 7.8) | 30 m*M* | 1.5 mL |
| 4.0 *M* | Potassium acetate | 100 m*M* | 1.25 mL |
| 1.0 *M* | Magnesium acetate | 2.7 m*M* | 135 µL |
| 100 m*M* | Spermidine trihydrochloride | 0.4 m*M* | 200 µL |
| 1.0 *M* | Dithiothreitol | 2.5 m*M* | 125 µL |
| 2.5 m*M* each | ${}^{15}$N-labeled amino acids | 0.3 m*M* | 6.0 mL |
| 100 m*M* | ATP | 1.2 m*M* | 600 µL |
| 20 m*M* | GTP | 0.25 m*M* | 625 µL |
| 500 m*M* | Creatine phosphate | 16 m*M* | 1.6 mL |
| 0.5% | Sodium azide | 0.005% | 500 µL |
| | RNase-free water | | 37.5 mL |
| | | Total | 50.0 mL |

sis buffer. All the other procedures are the same as those written in **Subheading 3.3.1.** (*see* **Note 7**).

3. To label the other amino acids selectively, follow the procedures written in **Subheading 3.3.1.** (*see* **Note 7**).

### 3.3.3. Analysis of Synthesized Proteins

1. Recover the translation solution with a long and narrow plastic pipet. Take 5 µL of the protein synthesis solution. Centrifuge it (20,000*g*; 5 min) and retain both supernatant and pellet.
2. Dissolve the pellet with 5 µL of water. Load both the supernatant and pellet fractions onto a 15–25% gradient sodium dodecyl sulfate polyacrylamide gel electrophoresis (SDS-PAGE) gel and visualize proteins by Coomassie Blue staining (*see* **Fig. 2**). Verify that the yeast ubiquitin is in the supernatant fraction.
3. Estimate the concentration of the synthesized yeast ubiquitin by comparing the band in SDS-PAGE gel with those of molecular weight markers.
4. The synthesized solution can be stored for 1 wk at 4°C or for several months at –20°C.

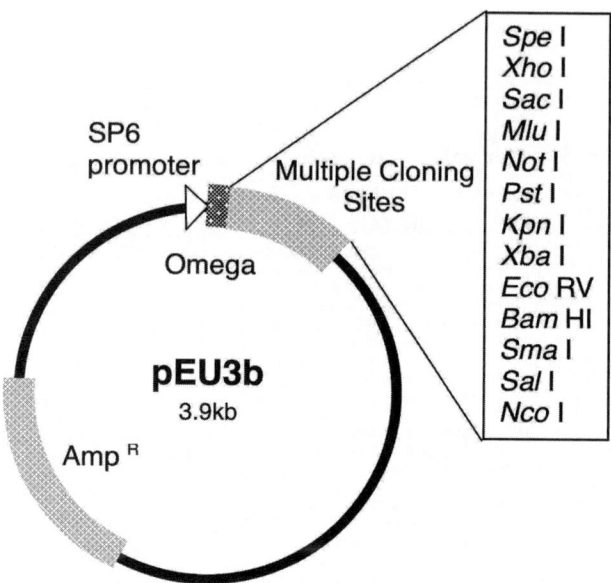

Fig. 1. Schematic drawing of the pEU3b cell-free expression plasmid.

## 3.4. NMR Sample Preparation of Yeast Ubiquitin Synthesized by Wheat Germ Cell-Free System

1. Put 1 mL of protein synthesis solution into Centricon YM-3 and centrifuge (5000*g*) at 4°C.
2. Centrifuge until the solution is concentrated to the volume of 200 μL (a few hours).
3. Equilibrate two MicroSpin G-25 columns with NMR sample buffer. At first, resuspend the resin in the column by vortexing gently.
4. Snap off the bottom closure and place the column in 1.5-mL microcentrifuge tubes. Then, prespin the column for 1 min at 735*g*.
5. Discard the solution in the microcentrifuge tubes and apply 250 μL of NMR sample buffer to each column. Spin the columns for 1 min at 735g and discard the buffer in the microcentrifuge tubes. Repeat this procedure eight times. Alternatively, if available, a MicroPlex 24 Vacuum (Amersham Biosciences) system may be useful to equilibrate MicroSpin G-25 Columns.
6. Place the columns in new 1.5-mL microcentrifuge tubes. Apply 100 μL of the synthesized protein to each tube.

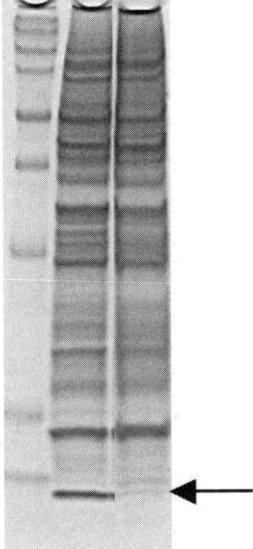

Fig. 2. SDS-PAGE analysis of [15]N labeled yeast ubiquitin synthesized by wheat germ cell-free protein synthesis system. Left lane; molecular weight markers. Middle lane; yeast ubiquitin translation solution. Right lane; wheat germ alone (control). Yeast ubiquitin is indicated by an arrow.

7. Spin the columns for 2 min at 735$g$. Collect the NMR sample from the bottom of the microcentrifuge tubes (*see* **Note 8**). Adjust the sample volume to 225 µL with NMR sample buffer. Add 25 µL of 10 m*M* DSS (chemical shift internal standard) dissolved in D$_2$O.
8. Alternatively, buffer exchange can be done by repeated dilution with NMR sample buffer and concentration using Centricon YM-3. For example, four repeats of fivefold dilution and concentration with NMR sample are sufficient for buffer exchange.

## 4. Notes

1. The pEU3b vector is utilized very effectively in producing large amount of proteins with wheat germ system (*6*). Another cell-free expression vector, pEU3-NII, is also commercially available (included in the cell-free kit) from TOYOBO (Osaka, Japan) and is based on the T7 transcription system.

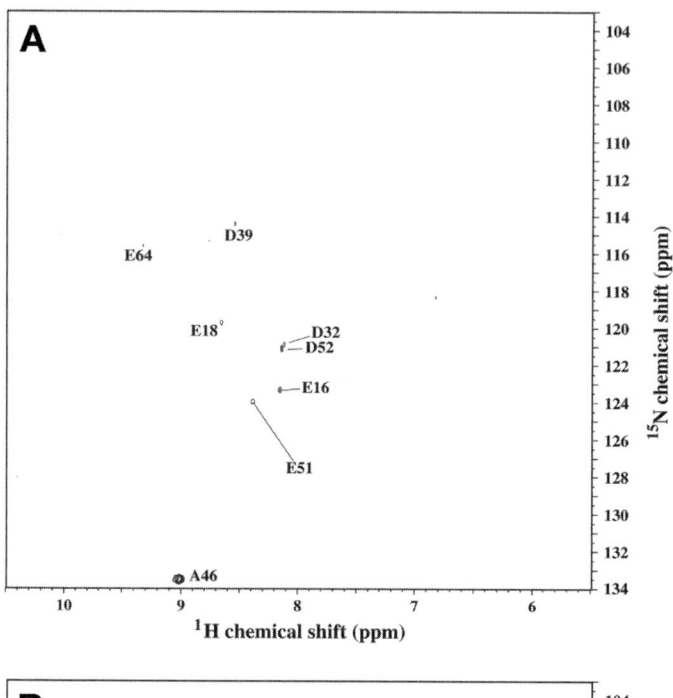

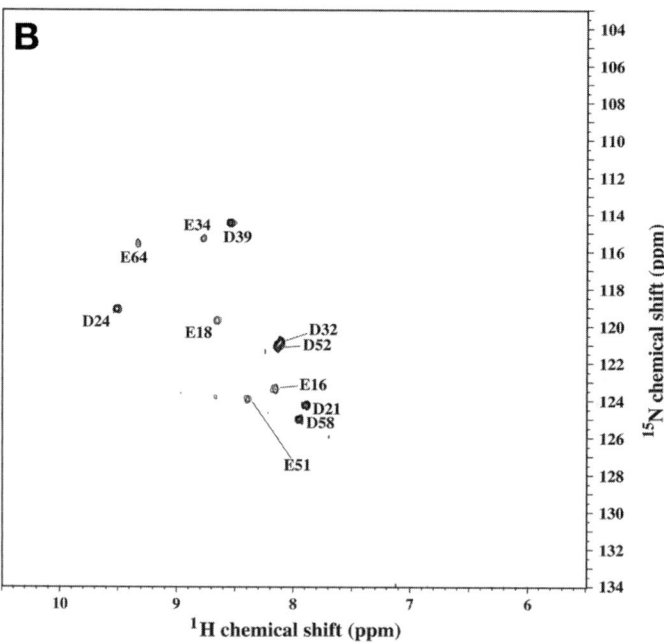

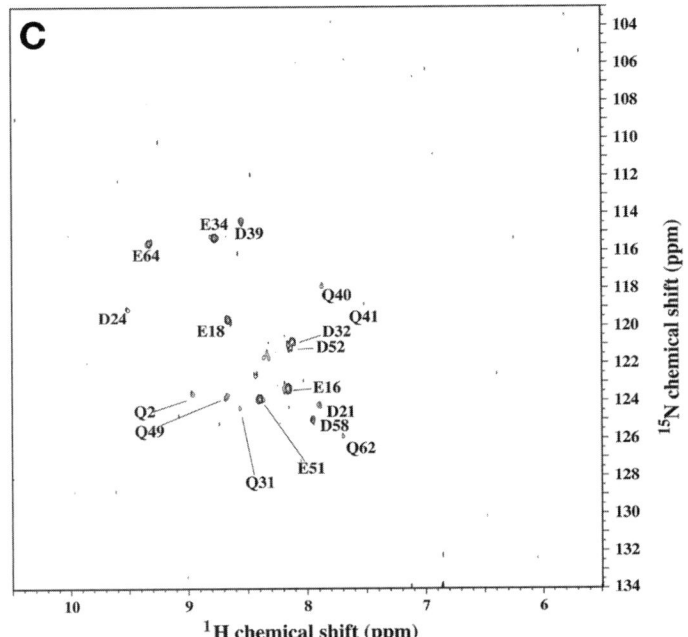

Fig. 3. $^{1}$H-$^{15}$N HSQC spectra of $^{15}$N-Ala (**A**), $^{15}$N-Asp (**B**), and $^{15}$N-Glu (**C**) selectively labeled yeast ubiquitin synthesized by wheat germ cell-free protein synthesis system. The conditions for spectral measurements are same as that in **Fig. 4**.

2. Wheat germ extract is commercially available from Cell Free Science or from TOYOBO (PROTEIOS, Osaka, Japan) or from ZOEGENE Corporation. Wheat germ extract may be prepared according to the protocol described in Chapter 5.

3. To obtain large amounts of proteins, trace contaminations of ribonucleases should be avoided throughout the transcription and translation steps. Therefore, great care must be taken if plasmid purification kits are used for template DNA preparations. These kits always use a solution containing RNase A in the first step of plasmid purification.

4. SP6 RNA polymerase is usually used at 37°C. However, this enzyme is 30% more active at 40°C than at 37°C. Therefore, 40°C is used in this protocol.

5. During ethanol precipitation, the use of ammonium acetate instead of sodium acetate is recommended because sodium salts inhibit the translation reaction.

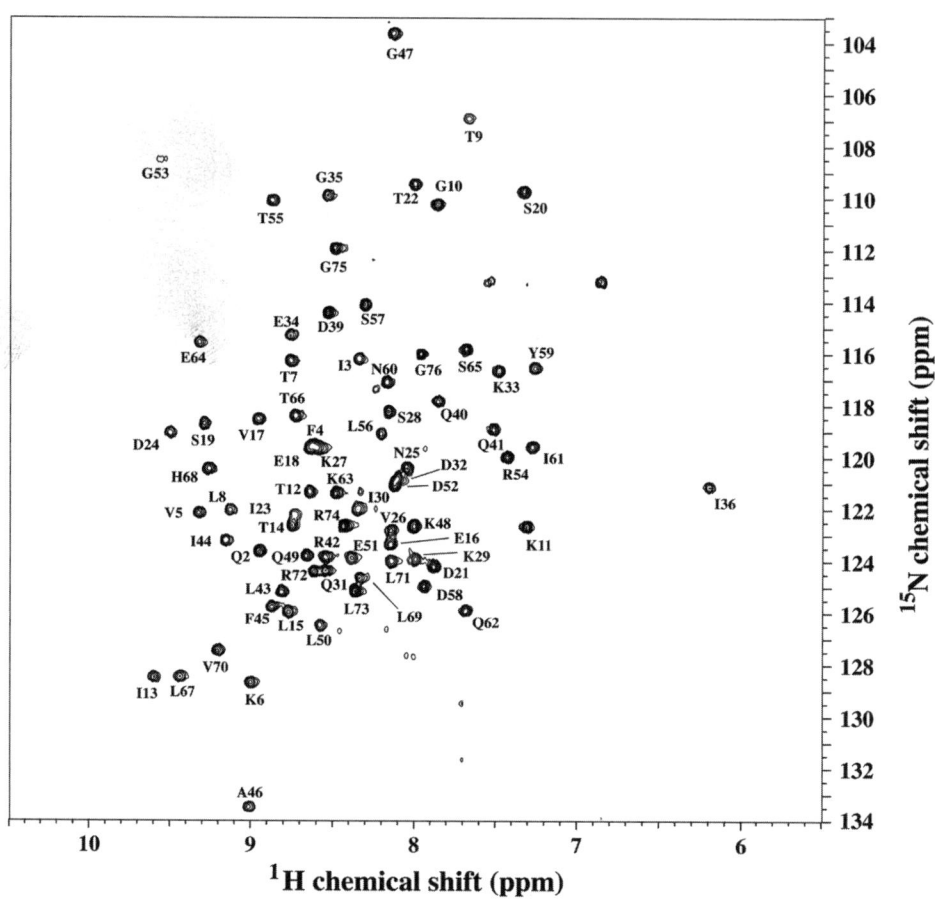

Fig. 4. ¹H-¹⁵N HSQC spectrum of nonpurified yeast ubiquitin synthesized by using the wheat germ cell-free protein synthesis system. All nuclear magnetic resonance (NMR) measurements were performed on a Bruker Avance500 spectrometer at 30°C. A 2D ¹H-¹⁵N HSQC spectrum *(11)* was acquired with 64 ($t_1$) × 1024 ($t_2$) complex points. The ¹H shifts were referenced to the methyl resonance of 4, 4-dimethyl-4-silapentane-1-sulfonic acid (DSS), used as an internal standard. The ¹⁵N chemical shifts were indirectly referenced using the ratio (0.101329118) of the zero-point frequencies to ¹H *(12)*. Spectral widths are 1600 and 6250 Hz in F1 and F2, respectively.

6. The translation reaction proceeds over 1 wk, as long as mRNA is supplied and the dialysis buffer is changed every 2 d. Should a high amount of protein be needed, continue the translation reaction. Some antibiotics may be added to the dialysis solution to avoid bacterial contaminations.

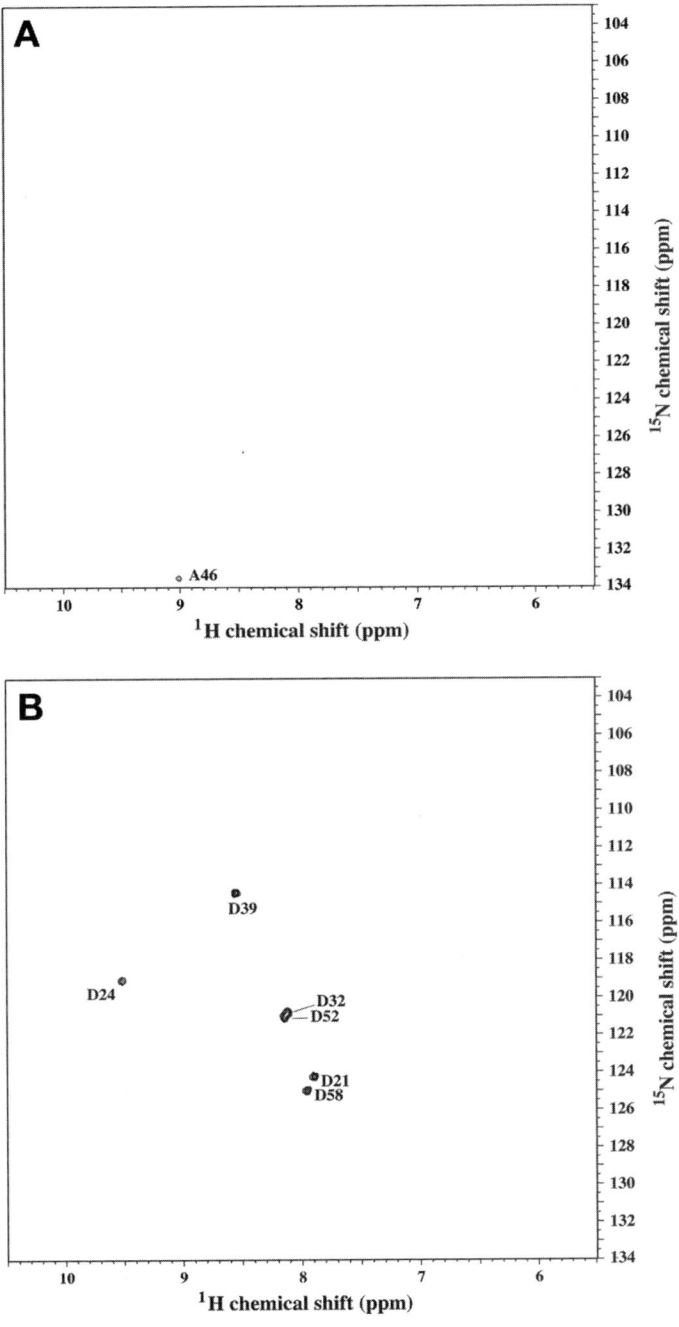

Fig. 5

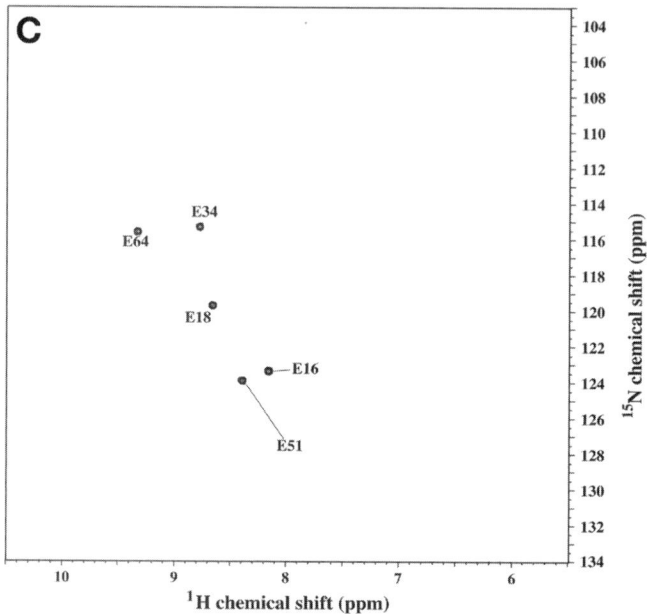

Fig. 5. *(continued from previous page)* ¹H-¹⁵N HSQC spectra of yeast ubiquitin synthesized with selectively labeled amino acid mixtures ([A] ¹⁵N-Ala, [B] ¹⁵N-Asp, and [C] ¹⁵N-Glu) in the presence of 1.0 m*M* aminooxyacetate (A,B) or in the presense of 1.0 m*M* aminooxyacetate and 0.1 mM L-methionine sulfoximine (C). The conditions for spectral measurements are same as that in **Fig. 4**.

7. Glutamine synthase and some transaminases are active in the wheat germ extract *(8)*. Therefore, selective labeling of Ala, Asp, and Glu requires the use of inhibitors (**Fig. 3**). In the case of Ala or Asp selective labeling, aminooxyacetate can be used to inhibit the activities of alanine transaminase and aspartate transaminase. In the case of Glu selective labeling, L-methionine sulfoximine can also be used to inhibit the activity of glutamine synthase. These inhibitors do not inhibit the protein synthesis activity of the wheat germ extract *(8)*.

8. White precipitation of proteins from wheat germ extract may be seen when the buffer of the translation solution is exchanged. Remove the precipitate by centrifuge (20,000*g*, 15 min).

9. **Figure 4** shows the ¹H-¹⁵N HSQC spectrum of nonpurified yeast ubiqutin prepared using the wheat germ in vitro system. All the signals visible in the **Fig. 4** match with those visible in the spectrum of purified

yeast ubiquitin prepared by *E. coli* cells (data not shown), and no other signals derived from some contaminations of other proteins are visible. This indicates that yeast ubiquitin prepared by wheat germ system assumes a correct fold, and no purification is necessary to measure the $^1$H-$^{15}$N HSQC spectrum of proteins prepared using this system. This feature may be useful to screen the "foldedness" *(1)* of proteins of interest very rapidly by NMR.

10. Under optimized conditions, amino acid metabolic enzymes may be completely inhibited (**Fig. 5**). Therefore, the wheat germ cell-free system may be used for amino acid-specific monitoring in the $^1$H-$^{15}$N HSQC spectra of proteins, which is one of the best ways for precise analysis of protein structures, and the intermolecular interactions between the proteins of interest and some ligands, in a high-throughput manner *(8)*.

## References

1. Montelione, G. T., Zheng, D., Huang, Y. J., Gunsalus, K. C., and Szyperski, T. (2000) Protein NMR spectroscopy in structural genomics. *Nat. Struct. Biol.* **7**, 982–985.
2. Yokoyama, S., Hirota, H., Kigawa, T., et al. (2000) Structural genomics projects in Japan. *Nat. Struct. Biol.* **7**, 943–945.
3. Wüthrich, K. (2000) Protein recognition by NMR. *Nat. Struct. Biol.* **7**, 188–189.
4. Shuker, S. B., Hajduk, P. J., Meadows, R. P., and Fesik, S. W. (1996) Discovering high-affinity ligands for proteins: SAR by NMR. *Science* **274**, 1531–1534.
5. Madin, K., Sawasaki, T., Ogasawara, T., and Endo, Y. (2000) A highly efficient and robust cell-free protein synthesis system prepared from wheat embryos: plants apparently contain a suicide system directed at ribosomes. *Proc. Natl. Acad. Sci. USA* **97**, 559–564.
6. Sawasaki, T., Ogasawara, T., Morishita, R., and Endo, Y. (2002) A cell-free protein synthesis system for high-throughput proteomics. *Proc. Natl. Acad. Sci. USA* **99**, 14,652–14,657.
7. Morita, E. H., Sawasaki, T., Tanaka, R., Endo, Y., and Kohno, T. (2003) A wheat germ cell-free system is a novel way to screen protein folding and function. *Protein Sci.* **12**, 1216–1221.
8. Morita, E. H., Shimizu, M., Ogasawara, T., Endo, Y., Tanaka, R., and Kohno, T. (2004) A novel way of amino acid-specific assignment in $^1$H-$^{15}$N HSQC spectra with a wheat germ cell-free protein synthesis system. *J. Biomol. NMR* **30**, 37–45.

9. Sambrook, J. and Russell, D. W. (2001) *Molecular Cloning, A Laboratory Manual, 3rd Ed.*, Cold Spring Harbor Laboratory Press, New York.

10. Kohno, T., Kusunoki, H., Sato, K., and Wakamatsu, K. (1998) A new general method for the biosynthesis of stable isotope-enriched peptides using a decahistidine-tagged ubiquitin fusion system: an application to the production of mastoparan-X uniformly enriched with [15]N and [15]N/[13]C. *J. Biomol. NMR* **12,** 109–121.

11. Schleucher, J., Schwendinger, M., Sattler, M., et al. (1994). A general enhancement scheme in heteronuclear multidimensional NMR employing pulsed field gradients. *J. Biomol. NMR* **4,** 301–306.

12. Wishart, D. S., Bigam, C. G., Yao, J., et al. (1995). [1]H, [13]C, [15]N chemical shift referencing in biomolecular NMR. *J. Biomol. NMR* **6,** 135–140.

# Index